Y0-EPE-317

DISCARDED

THE
UNIVERSITY OF WINNIPEG
PORTAGE & BALMORAL
WINNIPEG 2. MAN. CANADA

Molecular Orbital Methods in Organic Chemistry HMO and PMO

Studies in Organic Chemistry

Executive Editor

Paul G. Gassman
Department of Chemistry
The Ohio State University
Columbus, Ohio

Volume 1 Organoboranes in Organic Synthesis
Gordon M. L. Cragg

Volume 2 Molecular Orbital Methods in Organic Chemistry
HMO and PMO An Introduction
William B. Smith

Other Volumes in Preparation

QD
476
.S57

Molecular Orbital Methods in Organic Chemistry HMO and PMO

An Introduction

William B. Smith
Department of Chemistry
Texas Christian University
Forth Worth, Texas

MARCEL DEKKER, INC. New York 1974

Copyright © 1974 by Marcel Dekker, Inc.

ALL RIGHTS RESERVED.

Neither this book nor any part may be reproduced or transmitted in any form or by any means, electronic or mechanical, including photocopying, microfilming, and recording, or by any information storage and retrieval system, without permission in writing from the publisher.

MARCEL DEKKER, INC.
305 East 45th Street, New York, New York 10017

LIBRARY OF CONGRESS CATALOG CARD NUMBER: 73-90765

ISBN: 0-8247-6127-8

Current printing (last digit):
10 9 8 7 6 5 4 3 2 1

PRINTED IN THE UNITED STATES OF AMERICA

"------chemistry is a tiny perturbation on the face of nature------"

E. B. Wilson
The Robert A. Welch Foundation
Conference on Theoretical
Chemistry, November, 1972
Houston, Texas

CONTENTS

Preface		
Chapter 1	THE HYDROGEN ATOM AND HYDROGEN MOLECULE	1
	Introduction	1
	The Schrödinger Equation	2
	The Hydrogen Atom	5
	The Hydrogen Molecule	7
	Exercises	16
Chapter 2	HÜCKEL MOLECULAR ORBITAL THEORY	17
	Ethylene	18
	Butadiene	21
	Calculations of Coefficients	25
	Nonbonding MOs-Allyl	28
	Degenerate Orbitals-Cyclopropenyl	30
	Benzene	33
	Exercises	34
Chapter 3	BOND ORDERS, CHARGE DENSITIES, AND DIPOLE MOMENTS	35
	Bond Orders	35
	Charge Densities	38
	Dipole Moments	44
	Exercises	46
Chapter 4	HETEROATOM SYSTEMS	49
	Exercises	53

Chapter 5	THE PMO METHOD	55
	First-Order Perturbation Theory	56
	Internal Changes of α and β	60
	Amine Base Strengths	63
	The Union of π-Systems	68
	Exercises	73
Chapter 6	AROMATICITY, ANTIAROMATICITY, AND RESONANCE	75
	General Comments	75
	The PMO Methods, Localized Bonds, and Aromaticity	80
	The Hückel Rule	85
	Nonalternant Hydrocarbons	88
	Resonance Theory and MO-PMO Concepts	91
	Exercises	102
Chapter 7	CHEMICAL REACTIVITY	105
	The Nature of the Problem	105
	Electrophilic Substitution	107
	Heteroatom Systems	111
	Free Radical Substitution	117
	Nucleophilic Substitution	118
	Paralocalization Energies	119
	Ortholocalization Energy	121
	Carcinogenic Activity	124
	Exercises	126
Chapter 8	PERICYCLIC REACTIONS	127
	Evans' Rule, PMO Theory, and Pericyclic Reactions	130
	Anti-Hückel Systems	135
	Electrocyclic Reactions	138

Photochemical Reactions	141
Sigmatropic Rearrangements	143
Summary	146
The Conservation of Orbital Symmetry	146
Electrocyclic Reactions	152
Conclusion	154
Exercises	155
Bibliography	157
Index	159

PREFACE

The development of quantum mechanics during the early decades of this centruy led to the detailed insight of the nature of the chemical bond, molecular structure, and chemical reactions which we enjoy today. Freshman college textbookds incorporate atomic orbital pictures, utilize the aufbau principle, and present verbalizations of the covalent bond derived from quantum theory.

The use of large computers, the availability of computer programs and textbooks setting forth the latest and most sophisticated techniques have provided organic chemists with very powerful techniques indeed. But for the initiate and the experimentalist the fact remains that such complexities are largely of academic interest. We don't have the time, the facilities, or even the need to utilize these latest, most highly polished approximations to the solutions of the various quantum mechanical problems. For in the end, all problems of atomic structure require approximations to arrive at meaningful solution, and one must decide at what point one is willing to compromise between the labor required in making detailed calculations and simpler yet chemically useful answers derived by more extreme approximations.

The purpose of this text is to set forth a relatively simple but frequently useful approach to the properties and reactivities of organic molecules. The perturbation molecular orbital theory (PMO) is a spin-off of Hückel molecular orbital theory (HMO). It combines certain mathematical consequences from HMO treatment with intuitive insights into molecular constitution to provide useful results. For the purist, HMO theory contains such gross and conflicting approximations as to render it meaningless. However, the organic chemist is usually more of a pragmatist. He is willing to settle for useful results even though these may sometimes be due to a fortuitous cancellation of errors. As will be seen, both HMO and PMO theories provide answers for the chemist. If he will keep the limitations of the theories in mind, both theories can be of value.

This text covers a series of lectures given at TCU in part during the senior year's third semester of physical chemistry and in part in the physical organic chemistry course. As the title states, it is meant to be an introduction and not the definitive treatment of either HMO or PMO theories. Definitive treatments tend to be unusable as texts at the undergraduate level.

The author wishes to thank Dr. Clifford Venier and Dr. John E. Bloor for their careful reading of the manuscript and for their helpful discussion, and Mrs. Dorothy Clayton for her lovely typing.

William B. Smith

Molecular Orbital Methods in Organic Chemistry HMO and PMO

Chapter 1

THE HYDROGEN ATOM AND HYDROGEN MOLECULE

Introduction

The Bohr model of the hydrogen atom, introduced in 1913, represented a real break with the concepts of nineteenth century physics in that the electron moving about the nucleus was constrained to move in orbits in which the angular momentum was an integral multiple of $h/2\pi$. The h here is the ubiquitous Planck's constant, which had first appeared a few years earlier as a necessary proportionality constant in the equations accounting for black-body radiation and the photoelectric effect. When accompanied by the assumption that energy could only be absorbed or emitted by the atom during transitions between these allowed energy states, the quantization of angular momentum led immediately to a highly satisfactory calculation of the lines of the hydrogen spectrum.

The success of the Bohr theory as applied to hydrogen was so immediate and complete that many assumed the final link in atomic structure to have been achieved. Unfortunately, attempts to apply the model to other atoms quickly revealed shortcomings that required over a decade of work to correct.

Subsequent to the suggestion of de Broglie (1924) that high energy particles might have both wave and particle characteristics, Schrödinger combined a set of quantum conditions with the wave equations of classical mechanics to produce a new description of matter.

The Schrödinger Equation

The Schrödinger equation can be written in a deceptively simple form:

$$H\Psi = E\Psi \tag{1-1}$$

While a detailed discussion of the Schrödinger equation is out of place here, at least a superficial understanding will be helpful subsequently.

Mathematical instructions to perform some task are often conveyed in the form of operators. In the expression 3 x a, the symbol x is an operator which tells us to multiply a by 3. Similarly, in the expression

$$\frac{d}{dx} e^{ax} = ae^{ax} \tag{1-2}$$

the $\frac{d}{dx}$ symbol tells us to perform the differentiation. In the Schrödinger equation H is an operator, defined in more detail shortly, which tells us to perform a certain manipulation on the function Ψ.

The special form adopted in Eq. (1-2) is an example of a more general type of problem referred to as an eigenvalue problem. In eigenvalue problems some function, here the eigenfunction e^{ax}, undergoes an operation which results in the formation of the eigenfunction multiplied by a constant. The constant, a in (1-2), is referred to as the eigenvalue.

The Schrödinger equation is an eigenvalue problem. The H is an operator, Ψ is the eigenfunction, and the energy E is the eigenvalue. The eigenfunctions and eigenvalues which occur in the Schrödinger equation characterize atomic and molecular systems.

In classical particle physics it is often desirable to express the equation of motion by a function (the Hamiltonian) which embodies both the momenta and coordinates of the system. For time independent systems in classical motion H = KE (Kinetic energy) + V (potential energy). The Hamiltonian here is simply the total energy of the system.

Consider a particle moving in one dimension, say along the x axis. Then,

$$KE = \frac{1}{2} mv_x^2 = \frac{p_x^2}{2m} \qquad (1\text{-}3)$$

where p_x is the momentum along the axis. The total energy along the axis is

$$\frac{p_x^2}{2m} + V = E \qquad (1\text{-}4)$$

This equation can be converted to the Schrödinger form by replacing the momentum with the operator $\frac{h}{2\pi i} \frac{\partial}{\partial x}$ which gives us

$$-\frac{h^2}{8\pi^2 m} \frac{\partial^2}{\partial x^2} + V = E \qquad (1\text{-}5)$$

The problem with (1-5) in this form is that it tells us to operate on something but not what that something shall be. The something is the function Ψ. Thus,

$$-\frac{h^2}{8\pi^2 m} \frac{\partial^2 \Psi}{\partial x^2} + V\Psi = E\Psi$$

which rearranges to the more commonly seen form for a particle moving in one dimension

$$\frac{\partial^2 \Psi}{\partial x^2} + \frac{8\pi^2 m}{h^2} (E - V)\Psi = 0 \qquad (1\text{-}6)$$

In three dimensions (1-6) becomes

$$\nabla^2 \Psi + \frac{8\pi^2 m}{h^2} (E - V)\Psi = 0 \qquad (1\text{-}7)$$

where ∇^2 is the Laplacian operator

$$\frac{\partial^2}{\partial x^2} + \frac{\partial^2}{\partial y^2} + \frac{\partial^2}{\partial z^2}$$

The Hydrogen Atom

Now for the hydrogen atom KE = $p^2/2m$ and V = e^2/r, which is the Coulomb potential for an electron of charge e moving at a distance r from a nucleus of charge e. Therefore,

$$\nabla^2\Psi + \frac{8\pi^2 m}{h^2}(E - \frac{e^2}{r})\Psi = 0 \qquad (1\text{-}8)$$

There are a large number of solutions for such a differential equation, but the physical nature of the problem imposes certain limitations. The physical interpretation of Ψ is that of a probability distribution such that the probability of finding the electron at a given point in space is given by Ψ^2. A more correct way of expressing this is to say that the probability of finding the electron in a certain small volume element $d\tau$ (dxdydz in Cartesian coordinates) is $\Psi^2 d\tau$. Since the probability of finding the electron somewhere in space is 1, the value of the integral $\int\Psi^2 d\tau$ taken over all space must be 1. Wave functions that have this property are said to be normalized. For an unnormalized function, $\int\Psi^2 d\tau = N^2$ where N is a constant - the normalization factor. The normalized wave function in the latter case becomes Ψ/N.

Since the probability of finding the electron at any point in space is real, the Ψ function must be finite. Similarly, since no sudden changes in probability are expected the function must be continuous, and the function must be single valued as there is only one probability of finding the electron at a given point in space.

One possible solution to (1-8) is given by

$$\Psi = e^{-kr}$$

This solution corresponds to the hydrogen 1s state, lowest energy state available to an electron in the hydrogen atom. Other solutions describe the 2s, 2p, etc., states.

Application of the Schrödinger equation to other atoms in the periodic table requires approximate methods since one is faced with the insoluble classical three-body problem. However, very good wave functions are now available that account well for most, if not all, atomic spectra. These approximate wave functions not only give the electronic energy associated with each state, but also describe that region of space in which the probability of finding the electron is high. This last result gives us the atomic orbital pictures which decorate so many texts. As drawn, these orbitals are usually shown as that region of space enclosing 90% of the electron density. Wave functions for atoms above hydrogen in the periodic table are usually generated by altering the constants in the hydrogen wave functions and so have the same general form.

The Hydrogen Molecule

In attempting to find wave functions which will satisfactorily account for the energy states found in the hydrogen molecule one is immediately faced with an insoluble problem regarding the equations of motion for four bodies - two protons and two electrons. A variety of approximate methods have been explored, but we will follow only the line that leads to our goal.

The electrons in molecules can be pictures as occupying orbitals in much the same sense as those associated with atoms. Each of such electrons is described by a Ψ function such that Ψ^2 gives the probability distribution. That volume of space in which the probability density is high is pictured as the molecular orbital in much the same way as in atoms. For each Ψ there is an associated energy which approximates to the negative of the ionization energy for the electron. Electrons are assigned to moleculear orbitals in much the same way as to atomic orbitals, and Hund's rule regarding electron pairing applies, i.e., electrons will only pair up when all orbitals of the same energy are at least singly occupied.

The next problem is how to approximate the wave functions for the molecular orbitals. The approximation used here is that of a linear combination of atomic orbitals (LCAO) which for hydrogen assumes the form

$$\Psi = C_1\psi_1 + C_2\psi_2$$

where ψ_1 and ψ_2 are hydrogen 1s wave functions. This form follows from the expectation that in the region of one of the atoms the electron will be strongly influenced by that atom and be described by a wave function similar to the atomic wave function. The coefficients in the LCAO must be adjusted to minimize the total energy of the molecule and, for this purpose, one makes use of the method of variations.

The energy term of the Schrödinger equation can be extracted by multiplying both sides by ψ and integrating over all coordinates involved

$$\Psi H \Psi = \Psi E \Psi$$

$$E = \frac{\int \Psi H \Psi d\tau}{\int \Psi^2 d\tau} \tag{1-9}$$

There are several reasons for this operation rather than simply settling for the more obvious device of dividing both sides by Ψ. The function $H\Psi/\Psi$ frequently has no satisfactory minimum property. Furthermore, Ψ can have a zero value making the previous expression go to infinity.

Substitution of the LCAO gives

$$E = \frac{\int (C_1\psi_1 + C_2\psi_2)H(C_1\psi_1 + C_2\psi_2)d\tau}{\int (C_1\psi_1 + C_2\psi_2)^2 d\tau} =$$

$$\frac{C_1{}^2\int \psi_1 H\psi_1 d\tau + 2C_1C_2\int \psi_1 H\psi_2 d\tau + C_2{}^2\int \psi_2 H\psi_2 d\tau}{C_1{}^2\int \psi_1{}^2 d\tau + 2C_1C_2\int \psi_1\psi_2 d\tau + C_2{}^2\int \psi_2{}^2 d\tau} \tag{1-10}$$

This relation can be compressed by the following type of shorthand:

$$H_{11} = \int \psi_1 H\psi_1 d\tau$$

The integrals H_{11} and H_{22} are called <u>Coulomb integrals</u> and, of course, are equal to each other in this problem. In the limit where the nuclei are far apart the Coulomb integral reduces to the expression for the electron energy in atomic hydrogen. However, at closer distances the integral represents the energy of an electron moving in an orbital about one atom in the presence of the second. Hence,

$$H_{12} = \int \psi_1 H \psi_2 d\tau = \int \psi_2 H \psi_1 d\tau \ ^{*}$$

The integral H_{12} expresses the energy of the electron moving under the influence of both nuclei and is known as the <u>bond</u> or <u>resonance</u> integral. The value of H_{12} is negative as is that for H_{11}.

$$S_{11} = \int \psi_1 \psi_1 d\tau$$

This integral appears to have no specific name. In the case of the hydrogen atom this function is the probability of finding the electron somewhere in space. The wave functions are selected in such a way that the value of integral is one. Such wave functions are said to be <u>normalized</u>. We will work only with normalized wave functions hereafter.

$$S_{12} = \int \psi_1 \psi_2 d\tau$$

The integral S_{12} is a measure of the overlap of the hydrogen atomic 1s orbitals in forming the new molecular orbitals. When the atomic orbitals

---

*This equality is proven in a number of texts on quantum mechanics.

are far apart the value of S_{12} is zero. As the orbitals approach each other the value of S_{12} increases. At the equilibrium nuclear distance for the hydrogen molecule S_{12} is 0.59. The integral may be zero under other conditions too, i.e., when considering the interaction of an s orbital and a p orbital in the same atom. When the value of S_{12} is zero the wave functions are said to orthogonal.

Placing these definitions in Eq. (1-10) leads to

$$E = \frac{C_1^2 H_{11} + 2C_1C_2H_{12} + C_2^2 H_{22}}{C_1^2 + 2C_1C_2S_{12} + C_2^2} \qquad (1\text{-}11)$$

In order to find those C's that will minimize the energy one now differentiates E with respect to C_1 and C_2

$$\frac{\partial E}{\partial C_1} = \frac{2C_1H_{11} + 2C_2H_{12}}{(C_1^2 + 2C_1C_2S_{12} + C_2^2)}$$

$$- \frac{C_1^2H_{11} + 2C_1C_2H_{12} + C_2^2H_{22}(2C_1 + 2C_2S_{12})}{(C_1^2 + 2C_1C_2S_{12} + C_2^2)}$$

$$= \frac{2C_1H_{11} + 2C_2H_{12} - E(2C_1 + 2C_2S_{22})}{(C_1^2 + 2C_1C_2S_{12} + C_2^2)} = 0$$

Therefore,

$$C_1(H_{11} - E) + C_2(H_{12} - E\,S_{12}) = 0 \qquad (1\text{-}12)$$

and correspondingly from the $\partial E/\partial C_2$,

$$C_1(H_{12} - E\,S_{12}) + C_2(H_{22} - E) = 0 \qquad (1\text{-}13)$$

Equations (1-12) and (1-13) are called the secular equations. For a nontrivial solution for these linear homogeneous equations (i.e., for a solution other than $C_1=C_2=0$), then we have the condition that the secular determinant must be zero.

$$\begin{vmatrix} H_{11} - E & H_{12} - E\,S_{12} \\ H_{12} - E\,S_{12} & H_{22} - E \end{vmatrix} = 0$$

This determinant is of the form

$$\begin{vmatrix} a_{11} & a_{12} \\ a_{21} & a_{22} \end{vmatrix} ,$$

and the solution can be written explicitly as $a_{11} \times a_{22} - a_{21} \times a_{12}$. It is left as an exercise to convince oneself that expansion of the determinant and solution of the resulting quadratic in E gives

$$E = \frac{H_{11} \pm H_{12}}{1 \pm S_{12}} \qquad (1\text{-}14)$$

The solutions of the secular determinant for the hydrogen molecule thus give two energy states, and both electrons will occupy the lower (or ground state) level.

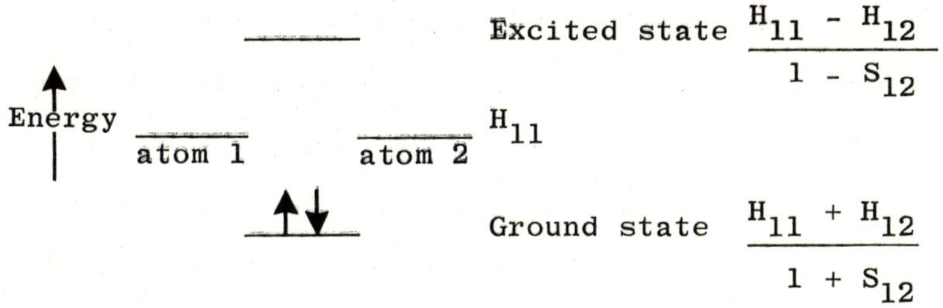

Because of the negative values associated with H_{11} and H_{12} the form of (1-14) with the positive signs corresponds to the lower state.

Calculations of the total energy of the molecule will require evaluation of the integrals defined above which, in turn, implies a knowledge of the Hamiltonian expressing all the attractive and repulsive interactions between the two electrons and the two protons. Happily, in carbocyclic systems it is usual to evaluate these integrals empirically.

For the hydrogen molecule ion (H_2^+) the problem and its solutions are formally as above. However, the equilibrium bond length will not be the same as for the neutral molecule and each of the integrals will assume a different value. This observation is made here to forewarn us that the assumption that these integrals will be the same for carbocyclic neutral and charged systems as well as for hetero-atom systems will prove to be an oversimplification.

Using the above values for E and the secular equations it is possible now to solve for the co-efficients. This will be demonstrated more explicitly in the next chapter. Substitution into the LCAO expression gives the two wave functions - one for each energy level:

$$\Psi_1 = \left(\frac{1}{\sqrt{2(1 - S_{12})}} \right) (\psi_1 + \psi_2)$$

$$\Psi_2 = \left(\frac{1}{\sqrt{2(1 - S_{12})}} \right) (\psi_1 - \psi_2)$$

Cross sections of these two wave functions may be sketched from the individual hydrogen atomic wave functions (showing only the radial portion of the wave function).

Ψ_1

Ψ_2

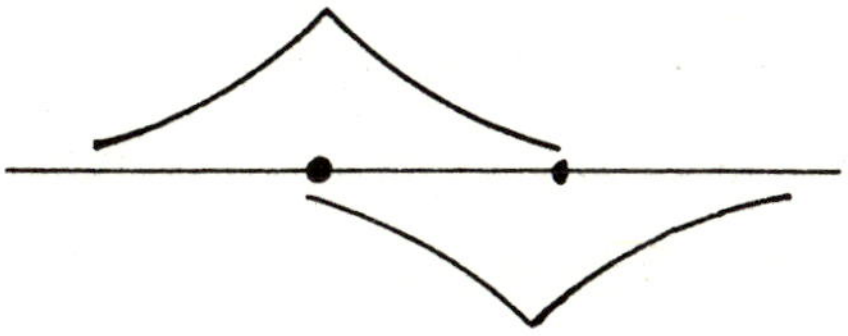

The electron density is given by the squares of the wave functions.

Ψ_1^2

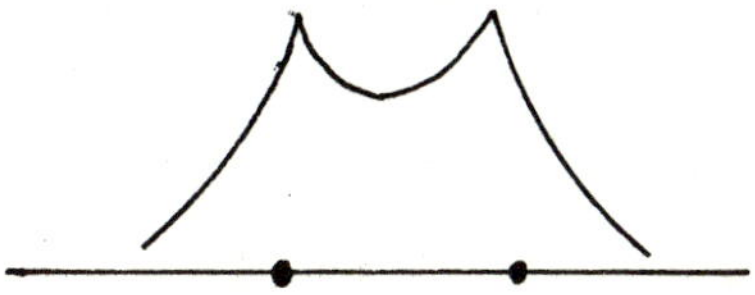

Ψ_2^2

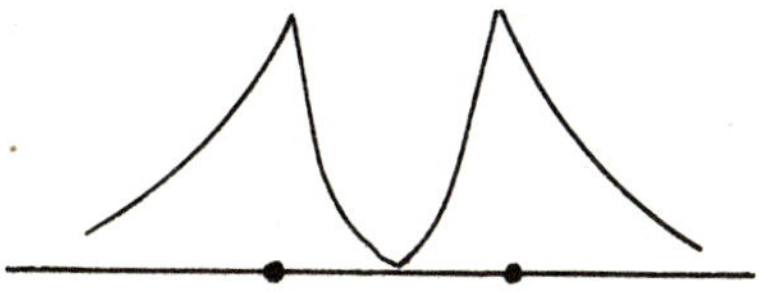

Since electron density piles up between the nuclei, the energy for Ψ_1 is less than for two individual atoms, and a bonding molecular orbital (MO) is produced. For ψ_2 at a point midway between the two atoms the probability ψ_2^2 is zero. This point is called a node. Molecular orbitals (MOs) that have a nodal point (or a plane in three dimensions) in between the nuclei are called antibonding MOs.

Exercises

1. Show that the solution $\psi = e^{-kr}$ leads to a value of $E = \frac{-2\pi^2 me^4}{h^2}$ from Eq. (1-8). Hint: It is necessary to make a change of variables. Also expressions such as $A + BX + CX^2 - - - = 0$ can be true for all values of X only if A, B, C, etc., are each equal to zero.

2. Given that

$$\frac{\partial^2\psi}{\partial x^2} + \frac{8\pi^2 m}{h^2}(E - 1/2kX^2)\psi = 0$$

with $\psi_0 = e^{-\alpha x}$ as a solution. Find α in terms of m and k; then find E_0 in the same terms. Another solution is given by $\psi_1 = xe^{-\alpha x^2}$. What is the relation of E_1 to E_0?

3. If ψ_1 and ψ_2 are each normalized and orthogonal (orthonormal), show that the LCAO wave functions below are orthogonal.

$$\Psi_1 = \frac{1}{\sqrt{2}}\psi_1 + \frac{1}{\sqrt{2}}\psi_2$$

$$\Psi_2 = \frac{1}{\sqrt{2}}\psi_1 - \frac{1}{\sqrt{2}}\psi_2$$

Chapter 2

HÜCKEL MOLECULAR ORBITAL THEORY

The method of variations and the LCAO approximation have been applied to many more complex systems than the hydrogen molecule. The development of large computers has made possible many sophisticated techniques of calculation. However, in 1931 Hückel introduced a rather simple set of approximations which have produced a surprisingly useful approach. The Hückel molecular orbital method (HMO) is often criticized because of the extreme nature of some of the approximations. However, if one bears in mind these limitations, one may still obtain useful insights into organic structures and reactions using the HMO approach.

Hückel's method applies to planar conjugated systems and depends immediately on the assumption that σ- and π-electrons in bonds act independently. While this assumption seriously affects calculations of electron spin resonance phenomena, it seemingly does not greatly influence many calculations of interest.

The Coulomb integrals between sp^2 hybridized carbons are assumed to be equal and to approximate the energy of an electron in a carbon 2p orbital. Hereafter, the Coulomb integral will be designated as α and taken as having a negative value.

The resonance integral, called β hereafter, and the overlap integral should obviously be a function of the length of the bond involved. In the HMO method it is usual to assume all β's are the same, although the variation of β with bond length can be incorporated if one wishes. Resonance integrals between nonadjacent carbons are set at zero. Moreover, in HMO calculations the obvious oversimplification is made that overlap integrals are zero. For desk-type calculations this greatly simplifies the mathematics and produces results not grossly different than those including reasonable values for S. In computer-based calculations S values are often incorporated.

Ethylene

The problem of the π-bond in ethylene is formally similar to the hydrogen molecule problem in that two atomic orbitals (in this case carbon

2p orbitals) are used in forming two new molecular orbitals. The secular determinant for hydrogen (p. 12) can be simplified by the inclusion of the HMO assumptions and our notation above to

$$\begin{vmatrix} \alpha - E & \beta \\ \beta & \alpha - E \end{vmatrix} = 0 \qquad (2\text{-}1)$$

Dividing both rows by β and setting x equal to $\alpha - E / \beta$ gives an even simpler form

$$\begin{vmatrix} x & 1 \\ 1 & x \end{vmatrix} = 0 \qquad (2\text{-}2)$$

$$x^2 - 1 = 0$$

$$x = \pm 1$$

$$E = \alpha \pm \beta \qquad (2\text{-}3)$$

Again bearing in mind that α and β are negative entities, it is seen that $\alpha + \beta$ represents the ground state energy of the ethylene π- system.

Level	Energy
___	$\underline{\alpha} - \underline{\beta}$
↑↓	$\underline{\alpha} + \underline{\beta}$

total

$$E_\pi = 2\,\underline{\alpha} + 2\,\underline{\beta}$$

From the secular equations

$$C_1(\alpha - E) + C_2 \beta = 0$$

$$C_1 \beta + C_2 (\alpha - E) = 0$$

When $E = \alpha + \beta$

$$\frac{C_1}{C_2} = 1$$

and when the condition $\Sigma C_i^2 = 1$ is applied, it may be seen by inspection that $C_1 = C_2 = 1/\sqrt{2}$. Similarly, for the root $\alpha - \beta$, $C_1 = -C_2 = 1/\sqrt{2}$. The two wave functions for the bonding and antibonding π-electron states of ethylene are

$$\psi_1 = 1/\sqrt{2}\ \psi_1 + 1/\sqrt{2}\ \psi_2$$

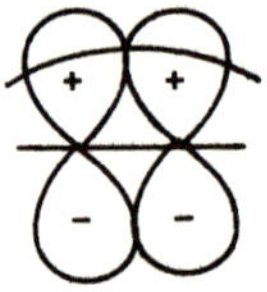

(2-4)

$$\psi_2 = 1/\sqrt{2}\ \psi_1 - 1/\sqrt{2}\ \psi_2$$

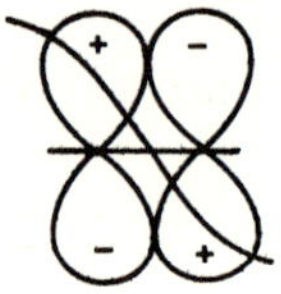

As shown, the antibonding function has a node between the carbons - a typical feature of antibonding states.

Butadiene

In butadiene the π-electrons occupy molecular orbitals made up of combinations of the four carbon 2p orbitals, i.e.,

$$\psi = C_1\psi_1 + C_2\psi_2 + C_3\psi_3 + C_4\psi_4 \qquad (2\text{-}5)$$

Applying the HMO postulates and the abbreviations used above one can write

$$\begin{vmatrix} \alpha-E & \beta & 0 & 0 \\ \beta & \alpha-E & \beta & 0 \\ 0 & \beta & \alpha-E & \beta \\ 0 & 0 & \beta & \alpha-E \end{vmatrix} = \begin{vmatrix} x & 1 & 0 & 0 \\ 1 & x & 1 & 0 \\ 0 & 1 & x & 1 \\ 0 & 0 & 1 & x \end{vmatrix} = 0$$

The latter determinant may be reduced by the method of minors as follows:

$$x \begin{vmatrix} x & 1 & 0 \\ 1 & x & 1 \\ 0 & 1 & x \end{vmatrix} - \begin{vmatrix} 1 & 1 & 0 \\ 0 & x & 1 \\ 0 & 1 & x \end{vmatrix} = 0$$

$$x^2 \begin{vmatrix} x & 1 \\ 1 & x \end{vmatrix} - x \begin{vmatrix} 1 & 1 \\ 0 & x \end{vmatrix} - \begin{vmatrix} x & 1 \\ 1 & x \end{vmatrix} + \begin{vmatrix} 0 & 1 \\ 0 & x \end{vmatrix} = 0$$

$$x^2 (x^2 - 1) - x (x - 0) - (x^2 - 1) = 0$$

$$x^4 - 3x^2 + 1 = 0$$

Faced with this fourth-order polynomial several devices may be tried to arrive at the four roots. If the equation can be reduced to a quadratic, the quadratic formula may be applied. Because of the nature of the bonding of carbon in π-systems, all roots must lie between $\pm$ 3. One can try 0 and all integers readily enough. Frequently, the equation will factor too. Finally, one may assign values to x and plot the curve to determine the roots. It is helpful to remember that the number of sign alterations between successive coefficients equals the number of negative roots. In our equation there are two such alterations. Consequently, there must be two negative roots.

The equation can be cast, in this case, in the form of a quadratic by setting $y = x^2$.

$$y^2 - 3y + 1 = 0$$

$$y = \frac{-b \pm \sqrt{b^2 - 4ac}}{2a}$$

Therefore,

$$y = \frac{3 \pm \sqrt{9 - 4}}{2}$$

$$y = 0.38,\ 2.62$$

$$x = \pm\ 0.62 \text{ and } \pm\ 1.62$$

Thus, the values of E in ascending order of energy are as shown below. The four π-electrons fill only orbitals one and two.

level	energy
____	$\alpha - 1.62\ \beta$
____	$\alpha - 0.62\ \beta$
↑↓	$\alpha + 0.62\ \beta$
↑↓	$\alpha + 1.62\ \beta$

$$E_{\pi}^{\text{total}} = 4\,\alpha + 4.48\,\beta$$

In classical butadiene the π-electrons are localized in bonds between C-1, C-2, and C-3, C-4, respectively. Since no interaction occurs between the C-2 and C-3 2p orbitals the value of β goes to zero for the 2-3 π-bond. The secular determinant is altered to

$$\left|\begin{array}{cc:cc} x & 1 & 0 & 0 \\ 1 & x & 0 & 0 \\ \hdashline 0 & 0 & x & 1 \\ 0 & 0 & 1 & x \end{array}\right| = 0$$

This determinant is in block diagonal form and factors directly to two 2 x 2 determinants identical to that for ethylene. The four roots then are $\alpha \pm \beta$ and $\alpha \pm \beta$. For this form of butadiene

$$E_{\pi}^{\text{total}} = 4\alpha + 4\beta$$

which is simply twice the energy of ethylene.

The energy difference between delocalized and classical (localized) butadiene is 0.48 β. This delocalization energy corresponds roughly to the concept of resonance energy as conceptualized by the organic chemist and is measured by heats of combustion or hydrogenation. Since β cannot be readily evaluated directly, it is generally assigned some experimental magnitude. The above treatment applied to benzene gives a delocalization energy (DE) of 2β. Since the experimental resonance energy for benzene is 36 k cal/mole, a value of 18 k cal/mole is convenient for β and gives fairly satisfactory agreement between experimental and HMO values for the DE in a number of aromatic hydrocarbons.

The calculated DE for butadiene on the above basis is 8.5 k cal/mole as compared to an experimental value of 3.5. Whether one considers this as good or bad agreement depends on one's point of view and consideration of the approximations inherent in the HMO method. The 18 k cal/mole value for β works best for aromatic hydrocarbons.

Calculations of Coefficients

Additional information about the butadiene molecule depends on a knowledge of the complete set of MO wave functions. The coefficients may be determined as before, i.e., by substituting each root, one at a time, into the secular equations and solving simultaneously. However, a convenient method is the so called "method of cofactors" illustrated below.

In this case it is possible to arrive at normalized coefficients using the relationships

$$N = \sqrt{\Sigma\ (C_n/C_1)^2} \qquad \text{n is the column}$$

$$\frac{C_n}{C_1} = + \frac{(\text{Cofactor})_n}{(\text{Cofactor})_1} \qquad \text{n is odd}$$

$$\frac{C_n}{C_1} = - \frac{(\text{Cofactor})_n}{(\text{Cofactor})_1} \qquad \text{n is even}$$

The ratio of cofactors (minors) for each coefficient must be set up first.

$$\frac{c_1}{c_1} = \frac{\begin{vmatrix} x & 1 & 0 \\ 1 & x & 1 \\ 0 & 1 & x \end{vmatrix}}{\begin{vmatrix} x & 1 & 0 \\ 1 & x & 1 \\ 0 & 1 & x \end{vmatrix}} = 1$$

$$\frac{c_2}{c_1} = \frac{\begin{vmatrix} 1 & 1 & 0 \\ 0 & x & 1 \\ 0 & 1 & x \end{vmatrix}}{\begin{vmatrix} x & 1 & 0 \\ 1 & x & 1 \\ 0 & 1 & x \end{vmatrix}} = \frac{1 - x^2}{x^3 - 2x}$$

$$\frac{c_3}{c_1} = \frac{\begin{vmatrix} 1 & x & 0 \\ 0 & 1 & 1 \\ 0 & 0 & x \end{vmatrix}}{\begin{vmatrix} x & 1 & 0 \\ 1 & x & 1 \\ 0 & 1 & x \end{vmatrix}} = \frac{1}{x^2 - 2}$$

$$\frac{c_4}{c_1} = \frac{\begin{vmatrix} 1 & x & 0 \\ 0 & 1 & 1 \\ 0 & 0 & x \end{vmatrix}}{\begin{vmatrix} x & 1 & 0 \\ 1 & x & 1 \\ 0 & 1 & x \end{vmatrix}} = \frac{-1}{x^3 - 2x}$$

For the root x = -1.62

n	C_n/C_1	$(C_n/C_1)^2$	C_n
1	1.00	1.00	0.372
2	1.62	2.62	0.602
3	1.62	2.62	0.602
4	1.00	1.00	0.372

$$\Sigma(C_n/C_1)^2 = 7.24$$

$$\sqrt{\Sigma(C_n/C_1)^2} = 2.69$$

Thus, the lowest energy wave function for butadiene is

$$\Psi_1 = 0.372\psi_1 + 0.602\psi_2 + 0.602\psi_3 + 0.372\psi_4$$

In similar fashion, the other roots of x can be shown to lead to

$$\Psi_2 = 0.602\psi_1 + 0.372\psi_2 - 0.372\psi_3 - 0.602\psi_4$$

$$\Psi_3 = 0.602\psi_1 - 0.372\psi_2 - 0.372\psi_3 + 0.602\psi_4$$

$$\Psi_4 = 0.372\psi_1 - 0.602\psi_2 + 0.602\psi_3 - 0.372\psi_4$$

These wave functions diagram as

Ψ_1 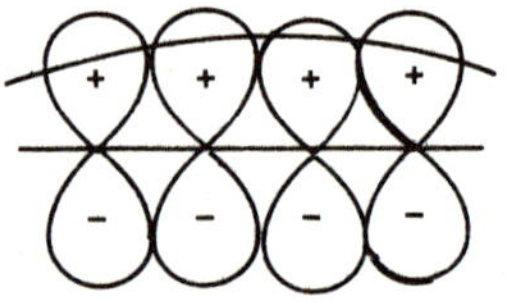

Ψ_2 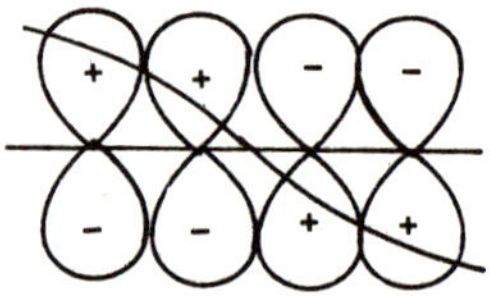

Ψ_3 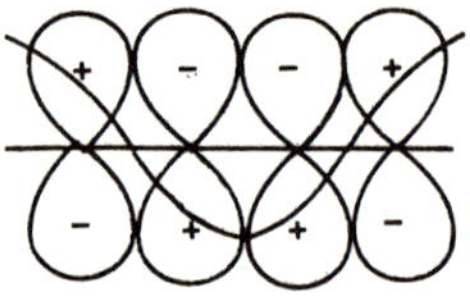

Ψ_4 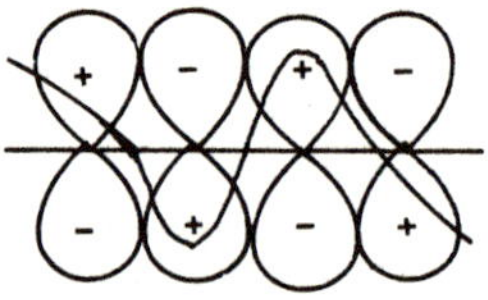

(2-5)

The wave function Ψ_1 is completely bonding, while Ψ_2 has one antibonding node. Bonding occurs between C-2 and C-3 in Ψ_3, but is antibonding elsewhere. Complete antibonding properties characterize Ψ_4.

Nonbonding MOs — Allyl

The three carbon π-system which is present in the allyl cation, free radical, and anion leads to the following secular determinant

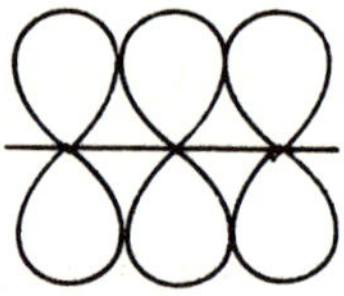

$$\begin{vmatrix} x & 1 & 0 \\ 1 & x & 1 \\ 0 & 1 & x \end{vmatrix} = 0 \qquad (2\text{-}6)$$

which leads in turn, to x = -2, 0, and +2. The

available energy levels are

Level	Energy
______	$\alpha - \beta$
______	α
______	$\alpha + \beta$

In the cation the first level is filled, while the radical contains these plus one electron in the second orbital, and the anion contains a full first level plus two electrons in the second orbital.

The wave functions can be shown to be

$$\Psi_1 = 1/2\ \psi_1 + 1/\sqrt{2}\ \psi_2 + 1/2\ \psi_3$$

$$\Psi_2 = 1/\sqrt{2}\ \psi_1 - 1/\sqrt{2}\ \psi_3$$

$$\Psi_3 = 1/2\ \psi_1 - 1/\sqrt{2}\ \psi_2 + 1/2\ \psi_3$$

Electrons which occupy the Ψ_2 orbital are characterized by an energy α, simply that of an electron localized in a carbon $2p$ orbital. Such electrons contribute nothing to the stability of the molecule. A diagram of Ψ_2 shows a node at C-2. Such

Ψ_2 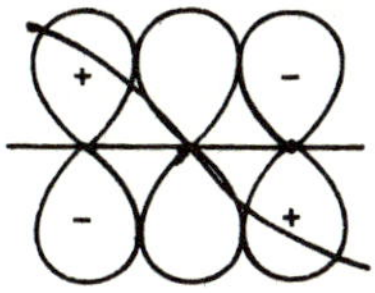 (2-7)

orbitals are called nonbonding molecular orbitals (NBMO), and will be discussed in considerable detail subsequently.

Degenerate Orbitals — Cyclopropenyl

The cyclopropenyl cation has been of interest because of the HMO predictions of a large delocalization energy. The secular determinant and roots for the system are as follows:

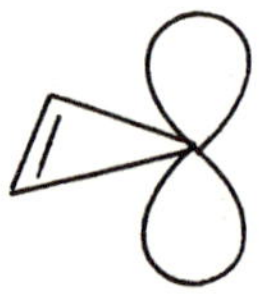

$$\begin{vmatrix} x & 1 & 1 \\ 1 & x & 1 \\ 1 & 1 & x \end{vmatrix} = 0 \qquad (2\text{-}8)$$

$$x = +1, \quad +1, \quad -2$$

$$E = \alpha + 2 \text{ , } \alpha - \beta, \ \alpha - \beta$$

The two orbitals of the same energy, $\alpha - \beta$, are said to be degenerate. In adding electrons to the system Hund's rule applies, and one electron would have to go into each orbital before doubling up in either can occur.

The determination of the coefficients for degenerate orbitals is more complex than for nondegenerate orbitals because there is an infinite variety of combinations of the atomic orbitals which may be used. Some of the tricks employed in meeting this situation are shown below.

For Ψ_1 the easiest procedure for this simple molecule is the method of elimination used for ethylene. The cofactor method could also be used.

$$x = -2$$

$$C_1x + C_2 + C_3 = 0$$

$$C_1 + C_2x + C_3 = 0$$

$$C_1 + C_2 + C_3x = 0$$

$$\Psi_1 = \frac{1}{\sqrt{3}}(\psi_1 + \psi_2 + \psi_3)$$

But x = 1 gives only $C_1 + C_2 + C_3 = 0$. One of two assumptions is usually made at this point, i.e., set one coefficient equal to zero or set two coefficients equal to each other. For Ψ_2 let us assume $C_1 = C_2$; then

$$C_1^2 + C_2^2 + C_3^2 = 1$$

$$C_1^2 + C_1^2 + (-2C_1)^2 = 1$$

$$6\,C_1^2 = 1$$

$$C_1 = 1/\sqrt{6}$$

and

$$\Psi_2 = 1/\sqrt{6}\,\psi_1 + 1/\sqrt{6}\,\psi_2 - 2/\sqrt{6}\,\psi_3$$

To find the coefficients of Ψ_3 we will construct a table of C-values and make use of the fact that the sum of all $C_1{}^2$'s squared equals one and so on for the C_2's and C_3's.

	C_1	C_2	C_3
Ψ_1	$1/\sqrt{3}$	$1/\sqrt{3}$	$1/\sqrt{3}$
Ψ_2	$1/\sqrt{6}$	$1/\sqrt{6}$	$-2/\sqrt{6}$
Ψ_3	a	b	c

$$(1/\sqrt{3})^2 + (1/\sqrt{6})^2 + (a)^2 = 1$$

$$a = 1/\sqrt{2}$$

$$(1/\sqrt{3})^2 + (1/\sqrt{6})^2 + (b)^2 = 1$$

$$b = -1/\sqrt{2}$$

$$(1/\sqrt{3})^2 + (-2/\sqrt{6})^2 + (c)^2 = 1$$

$$c = 0$$

$$\Psi_3 = 1/\sqrt{2}\ \psi_1 - 1/\sqrt{2}\ \psi_2$$

These MOs are diagramed as below.

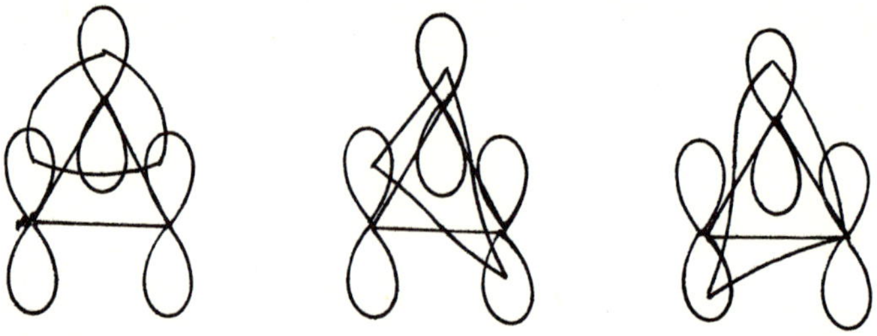

Ψ_1 Ψ_2 Ψ_3

(2-9)

Benzene

Because of the importance of benzene as the prime example of the aromatic hydrocarbons we give here the secular determinant, DE, and wave functions for the molecule

$$\begin{vmatrix} x & 1 & 0 & 0 & 0 & 1 \\ 1 & x & 1 & 0 & 0 & 0 \\ 0 & 1 & x & 1 & 0 & 0 \\ 0 & 0 & 1 & x & 1 & 0 \\ 0 & 0 & 0 & 1 & x & 1 \\ 1 & 0 & 0 & 0 & 1 & x \end{vmatrix} = 0$$

$$x = -2, -1, -1, 1, 1, 2$$

$$DE = (6\alpha + 8\beta) - (6\alpha + 6\beta) = 2\beta$$

$$\Psi_1 = 1/\sqrt{6}\ (\psi_1 + \psi_2 + \psi_3 + \psi_4 + \psi_5 + \psi_6)$$

$$\Psi_2 = 1/2\ (\psi_1 + \psi_2 - \psi_4 - \psi_5)$$

$$\Psi_3 = 1/\sqrt{12}\ (\psi_1 - \psi_2 - 2\psi_3 - \psi_4 + \psi_5 + 2\psi_6)$$

$$\Psi_4 = 1/2\ (\psi_1 - \psi_2 + \psi_4 - \psi_5)$$

$$\Psi_5 = 1/\sqrt{12}\ (\psi_1 + \psi_2 - 2\psi_3 + \psi_4 + \psi_5 - 2\psi_6)$$

$$\Psi_6 = 1/\sqrt{6}\ (\psi_1 - \psi_2 + \psi_3 - \psi_4 + \psi_5 - \psi_6)$$

Exercises

1. Write the secular determinant for methylene-cyclopropene, 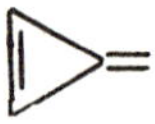 . Given the roots -2.170, -0.311, 1.00, and 1.481, calculate E_π and the DE.

2. How would the secular determinants of the following differ?

I

II

Given the roots below, calculate the DE for each. How does cross conjugation (II) affect the stability of the hexatriene system?

I	$x = \pm\ 1.802,\ \pm\ 1.247,\ \pm\ 0.445$
II	$x = \pm\ 1.932,\ \pm\ 1.000,\ \pm\ 0.518$

3. Write the secular determinant for cyclobutadiene, find the roots, and calculate the DE and the wave functions for the molecule. Diagram the wave functions.

Chapter 3

BOND ORDERS, CHARGE DENSITIES, AND DIPOLE MOMENTS

Bond Orders

For a number of years it has been realized that π-bonds in aromatic systems fell somewhere between single bonds and the ethylene double bond in character. Pauling suggested the weighted use of Kekule structures for benzene and other aromatics to assess the bond order, a single bond being zero and the ethylene π-bond being called one. On this basis the bond in benzene would be 0.5.

Coulson subsequently defined the "mobile bond order" or π-bond order as

$$p_{ij} = \sum_{n}^{occ} N\, C_{ni}\, C_{nj} \qquad (3\text{-}1)$$

where p_{ij} is the π-bond order between atoms i and j, N is the number of electrons in a given occupied orbital, and C_{ni} and C_{nj} are the appropriate normalized coefficients. The summation is taken over all occupied orbitals n.

For butadiene the bond orders are

$$p_{1,2} = 2(0.372)(0.602) + 2(0.602 \times 0.372) = 0.894$$

$$p_{2,3} = 2(0.602)(0.602) + 2(0.372)(-0.372) = 0.447$$

The results are in accord with our expectations that bonds C-1, C-2, and C-3, C-4 have more double bond character than C-2, C-3.

Since double bonds are shorter than single bonds, there have been numerous attempts to correlate bond orders with bond lengths. The following graph is based on ethylene (p 1.00, 1.33 Å), benzene (p 0.667, 1.39 Å), and the C-C single bond (p 0.00, 1.50 Å). The value for the single bond should be derived from a bond between sp^2 hydridized carbons rather than the value obtained from ethane (1.54 Å). Such a value is not directly obtainable, and values ranging from 1.47 to 1.52 Å have been set forth. The 1.50 Å value seems a reasonable compromise.

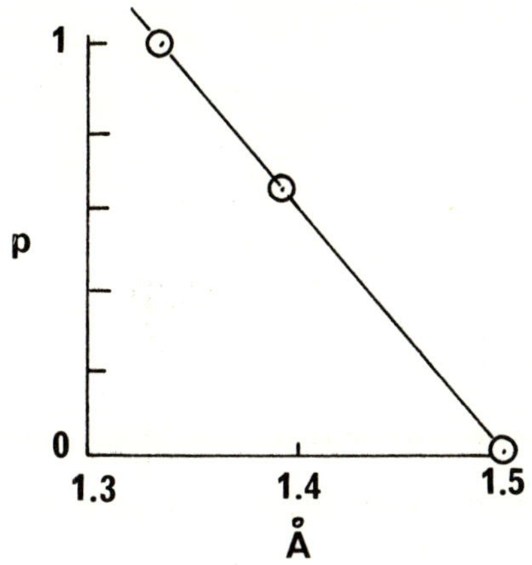

Using this graph, one estimates the bond lengths in butadiene as 1.36 and 1.45 Å, which are in good agreement with the experimental values of 1.37 and 1.47 Å. Good agreement for a large number of aromatic compounds is also found, so the above relation proves quite useful.

Now the total π-energy is simply

$$E_\pi = 2 \sum^{occ} \epsilon_n \tag{3-2}$$

and from the Schrödinger equation

$$\epsilon_n = \sum_i \sum_j C_{ni} C_{nj} H_{ij} \tag{3-3}$$

$$= \sum C_{ni}^2 \alpha + 2 \sum_i \sum_j^{i<j} C_{ni} C_{nj} \beta \tag{3-4}$$

Following our definition of bond order, Eq. (3-2) reduces to

$$E_{\pi} = N\alpha + 2\beta\ (\Sigma\ p_{ij}) \qquad (3\text{-}5)$$

For butadiene

$$E_{\pi} = 4\alpha + 2\beta\ (2 \times 0.894) + 0.447$$

$$= 4\alpha + 4.471\beta$$

<u>Charge Densities</u>

In ethylene and butadiene each carbon participates in three σ-bonds, and there is one π-electron for each p orbital in the π-system. Such carbons have an average π-electron density (ED) of 1.000.

In HMO theory the electron density is given by

$$ED_i = \overset{occ}{\Sigma}\ NC_i^{\,2} \qquad (3\text{-}6)$$

where the coefficients on atom i are summed over all occupied orbitals. For most purposes the deviation of ED from electrical neutrality is of more importance. This <u>charge density</u> is given by the general expression

$$q_i = K_{\pi i} - \overset{occ}{\Sigma}\ NC_i^{\,2} \qquad (3\text{-}7)$$

where $K_{\pi i}$ is the number of electrons donated by the atom in question to the π-system ($K_{\pi i}$ is one for carbon).

THE
UNIVERSITY OF WINNIPEG
PORTAGE & BALMORAL
WINNIPEG 2. MAN. CANADA

For butadiene

$$q_1 - q_4 = 1.000 - 2(0.372)^2 - 2(0.602)^2 = 0.000$$

$$q_2 = q_3 = 1.000 - 2(0.602)^2 - 2(0.372)^2 = 0.000$$

Similar results are obtained for ethylene and benzene.

It is convenient at this point to introduce the definitions of alternant (AHC) and nonalternant (non-AHC) hydrocarbons. The former are characterized by the fact that the atoms in the π-system may be divided into two sets, one starred and one unstarred, in such a fashion that no two starred or no two unstarred atoms are adjacent to each other. In even-AHCs the starred and unstarred positions can be interchanged to give a completely equivalent structure, while in odd-AHCs the two types of positions are not equivalent.

Even-AHCs

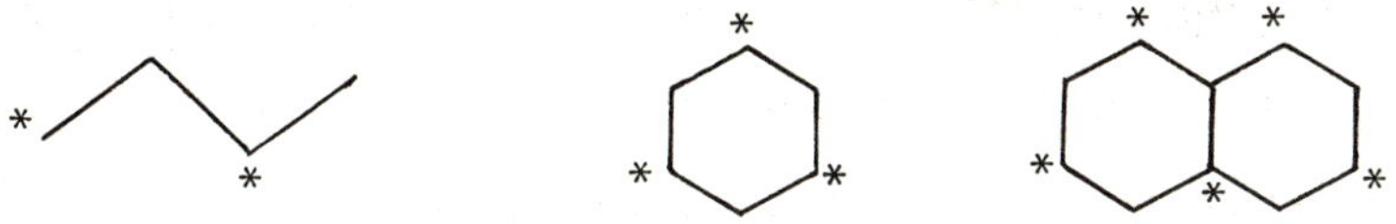

Odd-AHCs

Non-AHCs

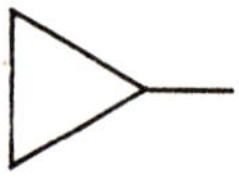 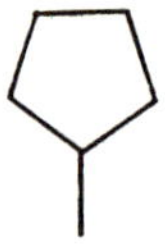 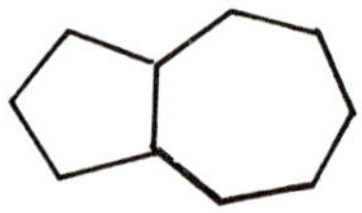

The above definitions are accompanied by some useful generalizations. As noted previously, the roots for butadiene and benzene occur in pairs symmetrically distributed about x = 0, and the same is true for all even-AHCs. For odd-AHCs there is an additional root at x = 0, which corresponds to the nonbonding molecular orbital (NBMO). One will also note that in "starring" the odd-AHCs we have deliberately chosen the pattern so that the starred positions outnumber the unstarred positions. The starred positions are referred to as "active" positions, and their importance as chemically reactive sites will become evident shortly.

Perusal of the odd-AHCs above, or any that can be written, quickly makes obvious the point that only reactive intermediates, i.e., radicals, anions, or cations, can exist in this form. Odd-AHC radicals are electrically neutral; the unpaired electron is in the NBMO. The importance of NBMO for odd-AHCs can be exemplified by the benzyl radical. As can be seen, the chemistry of benzyl is dependent upon the state of occupancy of the NBMO.

MO	Energy	Cation	Radical	Anion	
7	$\alpha - 2.1\beta$	___	___	___	
6	$\alpha - 1.3\beta$	___	___	___	Antibonding
5	$\alpha - \beta$	___	___	___	
4	α	___	↑	↑↓	Nonbonding
3	$\alpha + \beta$	↑↓	↑↓	↑↓	
2	$\alpha + 1.3\beta$	↑↓	↑↓	↑↓	Bonding
1	$\alpha + 2.1\beta$	↑↓	↑↓	↑↓	

For odd-AHCs it can be rigorously shown that the coefficients of the unstarred positions are zero for the NBMO. Furthermore, the algebraic sum of the coefficients on atoms adjacent to a given unstarred position is also zero. Thus, we have a quick method for calculating the NBMO coefficients.

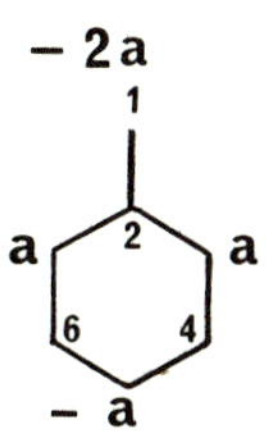

$$\Psi NBMO = C_1\Phi_1 + C_3\Phi_3 + C_5\Phi_5 + C_7\Phi_7$$

Arbitrarily, let $C_3 = a$; then $C_5 = -a$ and $C_7 = a$. Therefore, $C_3 + C_7 = 2a = -C_1$. Applying the normalization condition for the coefficients, $C_1^2 + C_3^2 + C_5^2 + C_7^2 = 1 = 7a^2$. Thus, $a = 1/\sqrt{7}$.

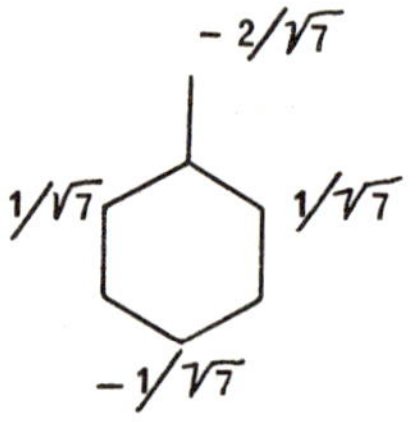

For the α-methylenenaphthalene NBMO the same technique may be applied.

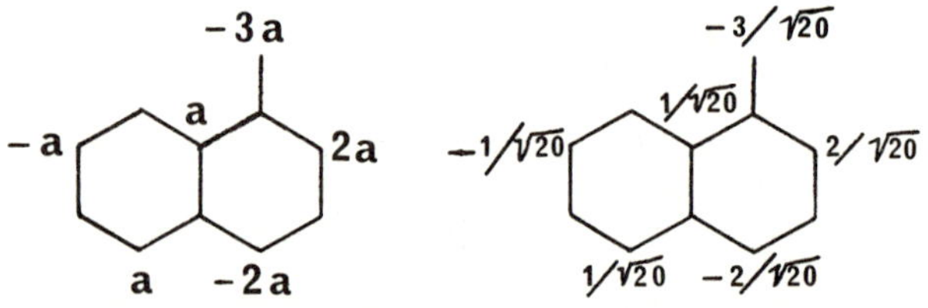

The a's were assigned by inspection starting at the ring junction. Here $20\ a^2 = 1$ and $a = 1/\sqrt{20}$.

The benzyl radical, of course, is electrically neutral having two electrons in each of the lower orbitals and one in the NBMO. Since the charge densities in the radical are zero at all positions, it is easily seen that the charge densities in the cation or anion will be simply the coefficients of the NBMO squared with the appropriate sign.

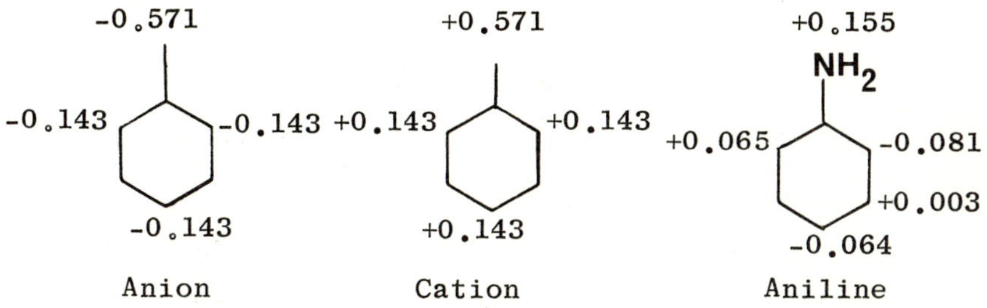

The Hückel MO values of the net charge density for aniline are shown for comparison. It is not surprising to find a similar charge distribution to the benzyl cation in nitrobenzene where the nitrogen bears a formal positive charge.

In the salt formation reactions of aromatic amines the amine nitrogen is protonated, removing the lone electron pair on nitrogen from interaction with π-system of the ring. The more deeply the lone pair becomes involved in conjugative interactions with the ring the more positive will be the charge at the nitrogen. The net π-charge at this position can be determined within the limits of PMO theory as proportional to those on the appropriate HCs.

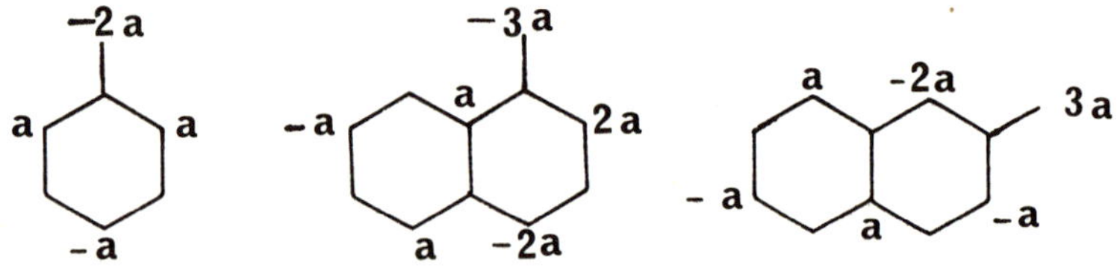

$a = 1/\sqrt{7}$ $\qquad$ $a = 1/\sqrt{20}$ $\qquad$ $a = 1/\sqrt{17}$

$1-4a^2 = +0.43$ $\qquad$ $1-9a^2 = +0.55$ $\qquad$ $1-9a^2 = +0.47$

The calculated order aniline > β-naphthyl > α-naphthyl corresponds with observed order of base strengths for the corresponding amines.

Dipole Moments

For non-AHCs the electron density will generally vary from unity. Therefore, non-AHCs will often have dipole moments due to π-charge disparities around the molecule. To assume no contribution to the molecular dipole moment as due to the underlying σ-framework is an obvious oversimplification. Clearly, polarization of the π-system will be reflected in the σ-system as well. But often the σ-contribution appears to be small, and dipole moments calculated from π-charges are at least "in the ball park." Remembering that dipole moments are vector quantities we may choose some arbitrary origin and calculate the contribution of each carbon to the total moment.

The charge densities for methylenecyclopropene are

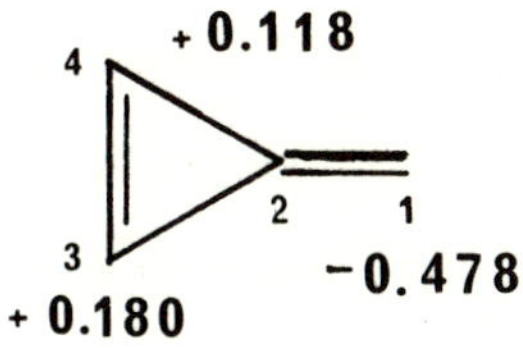

This example is chosen as the simplest non-AHC. Unfortunately, its dipole moment has not been experimentally measured. For simplicity we assume all bond lengths to be 1.40 Å. If we chose C-2 as the origin, then the contributing vectors are

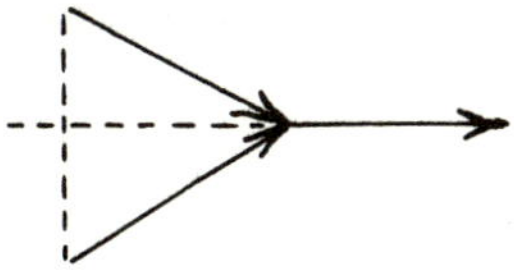

Now, since the definition of a dipole moment (μ) is the product of a distance vector (d) separating positive and negative charges (e), the component due to C-2 is nil (i.e., d = 0). The total contribution for the other three carbons is $\mu = 4.77$ $[(0.478)(1.40) + 2(0.180)(1.40)(\cos 30°)] = 5.26$ D. The factor, 4.77, is the appropriate conversion to Debye units.

Dipole moments calculated for non-AHCs by the simple HMO method are invariably too large. Experience with less severe approximations (and, therefore, more sophisticated calculations) shows the HMO method to always exaggerate the charge densities in non-AHCs with the resultant enhancement of the dipole moments in these compounds. Some idea of the magnitude of the error is given in Exercise 3 where results from two types of calculations are compared for fulvene.

Exercises

1. Calculate the bond orders and charge densities for cyclobutadiene using the wave functions developed in the previous chapter.

2. Given the roots and wave functions for the three lowest filled orbitals in fulvene, calculate the DE and values of q and p.

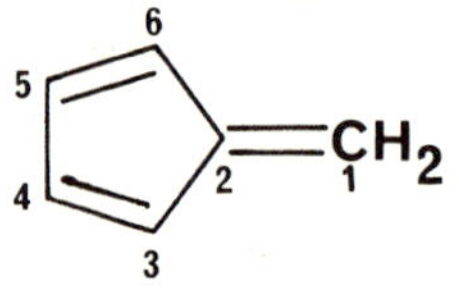

	ψ_1	ψ_2	ψ_3	ψ_4	ψ_5	ψ_6
Ψ_1	0.245	0.523	0.430	0.385	0.385	0.430

Ψ_2	0.500	0.500	_____	-0.500	-0.500	_____
Ψ_3	_____	_____	0.601	0.372	-0.372	-0.601

$x = -2.115, -1.000, -0.618$

3. The diagram below gives the dimensions of the fulvene molecule (derived from the x-ray structure of dimethylfulvene). Calculate the dipole moment using the HMO charge densities from Exercise 2 and the self consistent field values given. A value estimated from dimethylfulvene is 1.2, while a microwave determination gives an experimental value of 0.44.

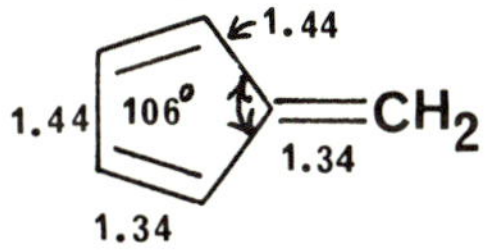

q SCF

1.	0.135	2.	0.008
3-6.	-0.030	4-5.	-0.042

4. Complete the calculation started in the text for the nonbonded wave functions and charge densities in the α- and β-naphthylmethyl anion (p. 44).

Chapter 4

HETEROATOM SYSTEMS

The binding energy for a carbon 2p electron is given by the Coulomb integral α. It goes without saying that the value of α will vary with the electronegativity of the carbon, and this, in turn, is affected by changes in hydridization or by substituents attached to the carbon. In the simplified HMO method previously considered all values of α in a carbon π-system were taken as being the same.

Similarly, the assumption is overly naive that all values of β for carbon-carbon bonds are the same in a given π-system. It is known that β varies with bond length, and various means of incorporating this variation have been attempted. Thus, two values of β should be used even for butadiene. However, the computational difficulties of introducing such variations quickly magnifies even HMO calculations to proportions requiring a computer.

When heteroatoms are introduced in a π-system new values of α and β are called for. Ideally one would like to calculate these from first principles, but more realistically one can derive empirical values more readily. This is done by expressing the values as a function of the values for carbon, i.e.,

$$\alpha_X = \alpha + h\,\beta$$

$$\beta_{CX} = k\,\beta$$

A variety of values of h and k for heteroatoms have appeared in the literature, and these have been reviewed by Streitweiser (see Bibliography). Because the values depend on whether the heteroatom contributes one or two electrons to the π-system, two sets of values for heteroatoms such as nitrogen and oxygen are usually given.

$$\alpha_N(1) = \alpha + 0.5\,\beta$$

$$\alpha_N(2) = \alpha + 1.5\,\beta$$

$$\beta_{C-N} = 0.8\,\beta$$

$$\beta_{C=N} = \beta$$

$$\alpha_O(1) = \alpha + \beta$$

$$\alpha_O(2) = \alpha + 2\ \beta$$

$$\beta_{C-O} = 0.8\ \beta$$

$$\beta_{C=O} = \beta$$

The secular determinant for methylene imine, CH_2=NH, will be analogous to that for ethylene.

$$\begin{vmatrix} \alpha - E & \beta_{CN} \\ \beta_{CN} & \alpha_N(1) - E \end{vmatrix} = \begin{vmatrix} \alpha - E & \beta \\ \beta & \alpha + 0.5 - E \end{vmatrix} = 0$$

Defining $(\alpha - E)/\beta = X$ gives

$$\begin{vmatrix} X & 1 \\ 1 & X + 0.5 \end{vmatrix} = 0$$

and

$$X^2 + 0.5X - 1 = 0$$

$$X = -1.28,\ 0.78$$

$$E = \alpha + 1.28\beta,\ \alpha - 0.78\beta$$

$$E_{\pi} = 2\alpha + 2.56\beta$$

which differs from the value of ethylene by 0.56 β.

It is now necessary to go back and determine wave functions, bond orders, and charge densities as before. Using the methods and definitions cited earlier one obtains:

$$\Psi_1 = 0.62\ \psi_1 + 0.78\ \psi_2$$

$$\Psi_2 = 0.78\ \psi_1 - 0.62\ \psi_2$$

$$p = 0.96$$

$$q_1 = +0.22,\ q_2 = -0.22$$

where (1) refers to the carbon and (2) to the nitrogen. Note that for ethylene $\Psi = 0.707\ \psi_1 \pm 0.707\ \psi_2$. The bond order, of course, is one, and there is no charge density. The bond order and charges for methylene imine are consistent with a resonance contributing structure such as

$$H_2\overset{+}{C} - \overset{-}{\ddot{N}}H$$

For pyridine it can be shown that the three occupied orbitals have energies

$$E_1 = \alpha + 2.11\ \beta$$

$$E_2 = \alpha + 1.17\ \beta$$

$$E_3 = \alpha + 1.00\ \beta$$

$$E_{total} = 6\ \alpha + 8.56\ \beta$$

The delocalization energy for pyridine is obtained by comparison with two ethylenes plus a methylene imine system, i.e., $(4\alpha + 4\beta) + (2\alpha + 2.56\beta) = 6\alpha + 6.56\beta$

$$DE = 2.00\ \beta$$

which is the same as the DE for benzene. The calculated charge densities for pyridine are given in the diagram. Here the q_N is still

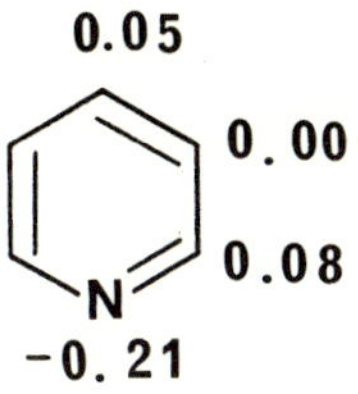

given by (1 - ED) since the nitrogen contributes only one electron to the π-system as do the carbons.

Exercises

1. Calculate the electron energy, wave functions, bond order, and charge densities for formaldehyde. Compare these entities with those for ethylene and methylene imine.

2. Show that -2.686, 0.186, and 1.000 are satisfactory roots for azacyclopropene. Determine the wave functions, bond orders, and charge densities for this system. Compare these with the cyclopropenyl anion (wave functions given on p. 31).

3. Given the following set of wave functions for phenol (filled orbitals only), calculate the electron densities and charge densities for all atoms.

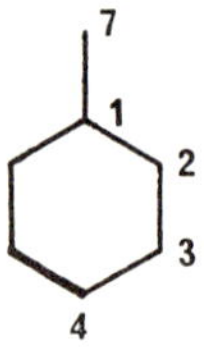

	Atom				
Wave functions	1	2	3	4	7
Ψ_1	0.603	0.312	0.181	0.144	0.597
Ψ_2	-0.173	0.175	0.456	0.563	-0.420
Ψ_3	0.000	0.500	0.500	0.000	0.000
Ψ_4	0.237	0.450	-0.083	-0.478	-0.545

Chapter 5

THE PMO METHOD

In HMO calculations the wave functions for the system are generated by the method of variations applied to a linear combination of the atomic orbitals which make up the system. The rèsultant wave functions describe the system subject to those limitations inherent in the HMO assumptions. While the availability of computers has allowed the application of HMO theory (and even more sophisticated treatments) to large π-electron systems, attempts have been made over the years to produce a shorthand type of calculation that would still provide valid insights into problems of molecular structure and chemical reactivity. The PMO method may be said to have been developed by Coulson, Longuet-Higgins, and Dewar (see Bibliography). It offers not only a rapid means of doing certain types of calculations but also some surprising knowledge about the meanings of resonance theory, the relative stabilities of hydrocarbons, the effects of heteroatoms and substituents, and the factors influencing certain types of chemical reactions.

As an introduction to the subject, consider a complex new sytem on which one would like to do calculations of the type covered in the preceding chapters. Suppose the wave functions for similar systems are already known. The solutions for the new system might be considered to arise from a perturbation of the latter. A case in point is the relation of the cyclopropenyl anion and azacyclopropene in the last chapter. Had we set the value of h in the Coulomb integral for nitrogen equal to zero the secular determinants for both molecules would have become the same. The wave functions for azacyclopropene reflect the perturbing affects of the inductive effect of the nitrogen, and this is reflected also in the energy levels and charges in the molecule.

First-Order Perturbation Theory

The following treatment of first-order perturbation theory provides much of the approach used to develop the PMO method and, thus, lays the background for what is to come. For the perturbed system let us write the Hamiltonian as

$$H = H^\circ + \lambda H^1$$

where H is the Hamiltonian for the unperturbed system and λH^1 is the perturbation. When the coefficient λ is zero the perturbation is nil. Let us define $\overset{\circ}{\Psi}_1$, $\overset{\circ}{\Psi}_2$, $\overset{\circ}{\Psi}_3$, ---- and $\overset{\circ}{E}_1$, $\overset{\circ}{E}_2$, $\overset{\circ}{E}_3$, ---- as eigenfunctions and eigenvalues of $\overset{\circ}{H}$. Furthermore, we will assume all eigenvalues to be different; the case with degeneracies requiring more specialized treatment. Similarly, Ψ_1, Ψ_2, Ψ_3, ---- and E_1, E_2, E_3, ---- characterize the values for the perturbed system. The corrections to the zeroth-order terms are Ψ_1^1, Ψ_2^1, Ψ_3^1, ---- and E_1^1, E_2^1, E_3^1, ----. Then,

$$\overset{\circ}{H}\overset{\circ}{\Psi}_i = \overset{\circ}{E}_i\overset{\circ}{\Psi}_i \tag{5-1}$$

$$\Psi_i = \overset{\circ}{\Psi}_i + \lambda\Psi_i^1 \tag{5-2}$$

$$E_i = \overset{\circ}{E}_i + \lambda E_i^1 \tag{5-3}$$

and

$$H\Psi_i = E_i\Psi_i \tag{5-4}$$

Substituting in (5-4)

$$(H^{\circ} + \lambda H^{1})(\psi_i^{\circ} + \lambda \Psi_i^{1}) = (E_i^{\circ} + \lambda E_i^{1})(\psi^{\circ} + \lambda \Psi_i^{1}) \quad (5\text{-}5)$$

$$H^{\circ} \Psi_i^{\circ} + \lambda H^{\circ} \Psi_i^{1} + \lambda H^{1} \Psi_i^{\circ} + \lambda^2 H^{1} \Psi_i^{1} = E_i^{\circ} \Psi_i^{\circ} + \lambda E_i^{\circ} \Psi_i^{1} + \lambda E_i^{1} \Psi_i^{\circ} + \lambda^2 E_i^{1} \Psi_i^{1} \quad (5\text{-}6)$$

Terms in λ^2 may be dropped from (5-6) since λ is a small fraction for most perturbations. Then subtracting (5-1) from (5-6) we get the first-order perturbation Eq. (5-7)

$$H^{\circ} \Psi_i^{1} + H^{1} \Psi_i^{\circ} = E_i^{\circ} \Psi_i^{1} + E_i^{1} \Psi_i^{\circ} \quad (5\text{-}7)$$

However, this equation still does not solve the problem as the corrective function Ψ^1 is not known. What is usually done is to approximate Ψ^1 as a linear combination of the unperturbed wave functions

$$\Psi_i^{1} = \Sigma a_{ij} \Psi_j^{\circ}$$

Substituting

$$H^{1} \Psi_i^{\circ} + \sum_j a_{ij} H^{\circ} \Psi_j^{\circ} = E_i^{1} \Psi_i^{\circ} + E_i^{\circ} \sum_j a_{ij} \Psi_j^{\circ} \quad (5\text{-}8)$$

Multiplying by Ψ_k° and integrating

$$\int \Psi_k^{\circ} H^1 \Psi_i^{\circ} d\tau + \Sigma a_{ij} \int \Psi_k^{\circ} H^{\circ} \Psi_j^{\circ} d\tau =$$

$$E_i^1 \int \Psi_k^{\circ} \Psi_i^{\circ} d\tau + E_i^{\circ} \Sigma a_{ij} \int \Psi_k^{\circ} \Psi_j^{\circ} d\tau \qquad (5\text{-}9)$$

$$\int \Psi_k^{\circ} H^1 \Psi_i^{\circ} d\tau + \sum_j a_{ij} E_j^{\circ} \delta_{jk} =$$

$$E_i^1 \int \Psi_k^{\circ} \Psi_i^{\circ} d\tau + E_i^{\circ} \sum_j a_{ij} \delta_{kj} \qquad (5\text{-}10)$$

When j = k

$$\int \Psi_k^{\circ} H^1 \Psi_i^{\circ} d\tau + a_{ik} E_k^{\circ} =$$

$$E_i^1 \delta_{ik} + E_i^{\circ} a_{ik} \qquad (5\text{-}11)$$

Now if i = k, (5-11) reduces to

$$\int \Psi_i^{\circ} H^1 \Psi_i^{\circ} d\tau = E_i^1 \qquad (5\text{-}12)$$

Thus, the perturbation correction to E_i does not depend on a knowledge of the perturbed wave function - only the perturbation operator.

When $i \neq k$ Eq. (5-11) reduces to

$$\int \Psi_k^{\circ} H^1 \Psi_i^{\circ} d\tau + a_{ik} (E_k^{\circ} - E_i^{\circ}) = 0 \qquad (5\text{-}13)$$

and

$$a_{ik} = \frac{\int \Psi_k^{\circ} H^1 \Psi_i^{\circ} d\tau}{(E_k^{\circ} - E_i^{\circ})} \tag{5-14}$$

The coefficients a_{ii} can not be determined by (5-14) since E_i° will equal E_k°. Having determined all other coefficients, however, normalization may be employed to determine a_{ii}.

The perturbed wave equation will be of the form

$$\Psi_i = \Psi_i^{\circ} + \sum_j \frac{\int \Psi_i^{\circ} H^1 \Psi_i^{\circ} d\tau}{(E_i - E_j^{\circ})} \Psi_j^{\circ} \tag{5-15}$$

Internal Changes of α and β

Let us consider what happens to the energies of a conjugated π-system when a heteroatom replaces a carbon. Rather than making the usual approximations of Chapter 4 and solving new sets of determinants a perturbational approach will be used. Furthermore, the definitions used above will be continued except for a convenient change in the subscripted variable to be consistent with our earlier expressions for the LCAO, i.e., for the u-th MO

$$\Psi_u = \sum_i C_{ui} \psi_i$$

Reiterating from above

$$E_u = E_u^{\circ} + \lambda E_u^1$$

and from (5-12)

$$E_u^1 = \int \Psi_u^{\circ} H^1 \Psi_u^{\circ} d\tau$$

For convenience λ will be taken inside H^1 leading to

$$\delta E_u = E_u^1 - E_u^{\circ} = \int \Psi_u^{\circ} H^1 \Psi_u^{\circ} d\tau \qquad (5\text{-}16)$$

From our earlier definitions of the Coulomb and resonance integrals of carbon atom i

$$\alpha_i = \int \psi_i H^{\circ} \psi_i d\tau$$

$$\beta_{ij} = \int \psi_i H^{\circ} \psi_j d\tau$$

and for the perturbed functions

$$A_i = \int \psi_i H \psi_i d\tau$$

$$B_{ij} = \int \psi_i H \psi_j d\tau$$

The correction terms then become

$$\delta \alpha_i = A_i - \alpha_i = \int \psi_i H^1 \psi_i d\tau$$

$$\delta \beta_{ij} = B_i - \beta_i = \int \psi_i H^1 \psi_j d\tau$$

Now (5-16) can be rewritten as

$$\delta E_u = \int (\Sigma\, C_{ui}\, \psi_i)\, H^1\, (\Sigma\, C_{ui}\, \psi_i)\, d\tau =$$

$$\Sigma C_{ui}^{\;2} \int \psi_i H^1 \psi_i\, d\tau +$$

$$\sum_{i \neq j}\sum C_{ui}\, C_{uj} \int \psi_i\, H^1\, \psi_j\, d\tau \qquad (5\text{-}17)$$

Inserting our definitions above

$$\delta E_u = \sum_i C_{ui}^{\;2}\, \delta\alpha_i + 2 \sum_{i<j}\sum C_{ui}\, C_{uj}\, \delta\, \beta_{ij} \qquad (5\text{-}18)$$

The factor of 2 appears in the last equation because $\delta\, \beta_{ij} = \delta\, \beta_{ji}$ (see p. 10).

For the total energy change due to alterations in α and β (each orbital containing two electrons)

$$\delta E = 2 \sum_u^{occ} E_u^1 - 2 \sum_u^{occ} E_u^{\circ} = 2 \sum_u^{occ} \delta E_u =$$

$$2 \sum^{occ} (\Sigma C_{ui}^{\;2}\, \delta\, \alpha_i + 2 \sum_{i<j}\sum C_{ui}\, C_{uj}\, \delta\, \beta_{ij}) \qquad (5\text{-}19)$$

From our previous definitions of the electron density ED and bond order p_{ij} (5-19) may be rewritten as

$$\delta\, E = \sum_i\, ED_i\, \delta\alpha_i + 2 \sum_{i<j}\sum p_{ij}\, \delta \quad {}_{ij} \qquad (5\text{-}20)$$

We will have several occasions to use this equation again, but we may immediately check one of the consequences of heteroatom substitution seen in the last chapter. If the approximation

$\beta_{CN} = \beta$ is again employed, (5-20) reduces to

$$\delta E = \sum_i ED_i \, \delta \alpha \qquad (5\text{-}21)$$

Since the ED in our AHC is unity and $\delta \alpha_i = h\alpha$, δE is simply $h\,\alpha$. The DE for a heterocycle such as pyridine will be the same as that of the corresponding hydrocarbon since the methylene imine fragment used in our model compound will also be reduced by $h\,\alpha$. Of course, this is in agreement with our HMO derived conclusions for the one specific case.

Amine Base Strengths

Previously (p. 43) we saw that a qualitative assessment of amine base strengths could be made on the assumption that a CH_2^- group could represent an NH_2 group. The equilibrium between two bases and their conjugate acids may be written as

$$B_1H^+ + B_2 \rightleftharpoons B_1 + B_2H^+$$

If we are concerned with base strengths in aqueous solution, then B_1H^+ would be a water molecule and B_1 the hydroxide ion. The standard free energy change for the reaction is given by

$$\Delta G^\circ = -RT \ln K = 2.303\, RT\, (pK_{a1} - pK_{a2})$$

In discussions of relative base strengths of a related series of amines it is often assumed that electrostatic interactions in solution may be neglected and that ionization entropies are about the same. If so, then the free energy change contains only potential energy terms related to the change in π-electron energies. Given B_1 as some reference base the above equation can be rewritten as

$$2.303\ RT\ pK_a = \text{constant} - (E_{BH^+} - E_B) \qquad (5\text{-}22)$$

Now for the monoaza substituted heterocycles such as pyridine, quinoline, and isoquinoline the total π-energy from (5-21) above is

$$E_B = E^{\circ} + h_N\ \beta$$

In the conjugate acid the parameter h_{N^+} will assume some value larger than h_N to express the enhanced electronegativity of the positively charged nitrogen

$$E_{BH^+} = E^{\circ} + h_{N^+}\ \beta$$

The change in π-energy on protonation then reduces to

$$(E_{BH^+} - E_B) = (h_{N^+} - h_N)\ \beta$$

Thus, all the monoaza heterocyclics are predicted by PMO theory to have the same base strength. In point of fact, data for a number of these compounds show pK_as in the range of 4.2 - 5.2. Acridine is an exception having a value of 5.6 in water.

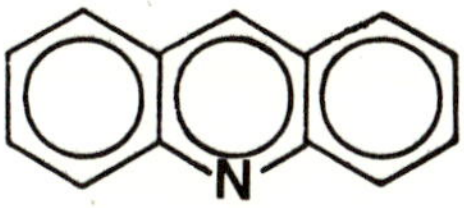

Acridine

The pK_a is comparable with the others in a 50% aqueous alcohol solvent.

Spectroscopic and nuclear magnetic resonance studies have shown that amine substituted heterocyclics such as the two below protonate on the ring nitrogen first.

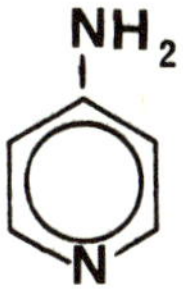

4-Aminopyridine

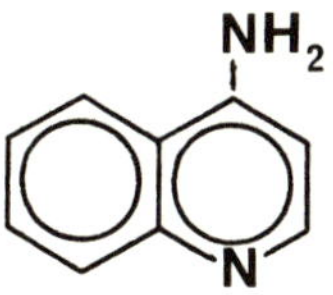

4-Aminoquinoline

The pK_as of a series of such compounds have been examined, and the range of values is considerably larger than for the heterocyclics above, the pK_a

being influenced by the presence and orientation of the amine group on the ring. The energy changes in the 4-aminoquinoline system may be summarized as follows [note the use of Eq. (5-21) implied in (b) and (d)].

ΔE_π

$h_N \beta$

$ED_4 \; h_N \; \beta$

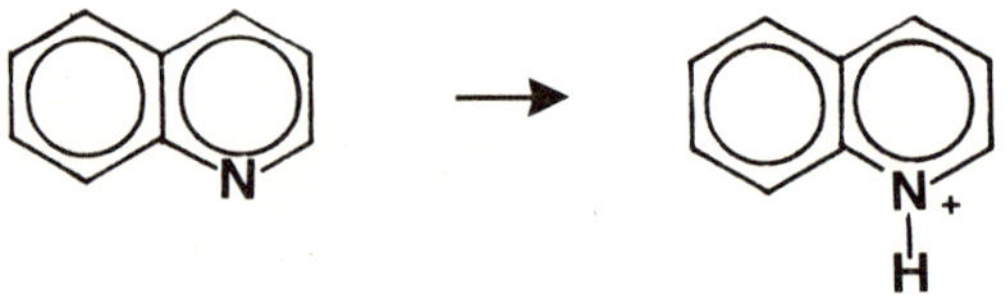

$(h_{N^+} - h_N) \beta$

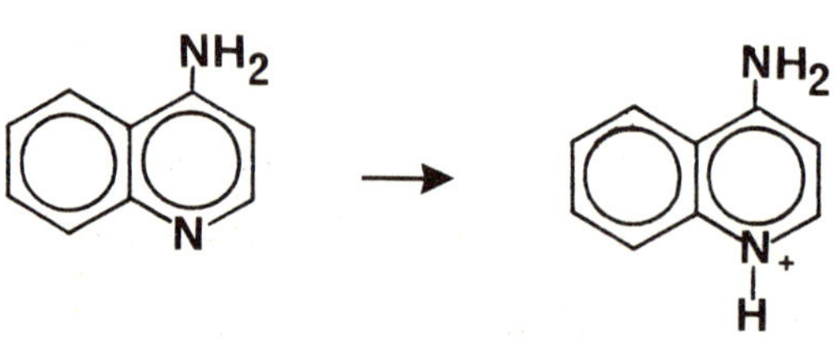

$ED_4 \; (h_{N^+} - h_N) \beta$

The electron density ED_4 can be obtained from the square of NBMO coefficients per p. 42-43 . From (5-22) above then

$$2.303\ RT\ pK_a = \text{constant} - ED_4\ (h_{N^+} - h_N)\ \beta \qquad (5\text{-}23)$$

Since ED_4 and $(h_{N^+} - h_N)$ are positive numbers while β is negative, an increase in ED_4 will lead to an increase in pK_a.

Longuet-Higgins (1950) has applied the above scheme to some twenty amine heterocycles. A least squares fit of his listed data gives a correlation coefficient of 0.911 (a perfect linear fit is 1; no fit is zero). The two poorest points on the curve are due to 1-aminoacridine and 9-aminophenanthridine.

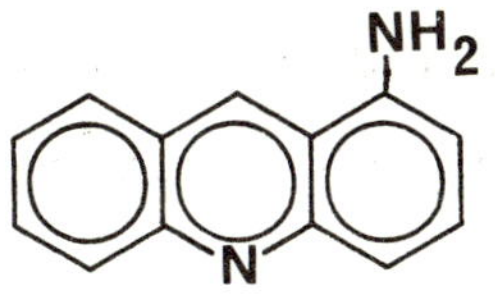

1-Aminoacridine

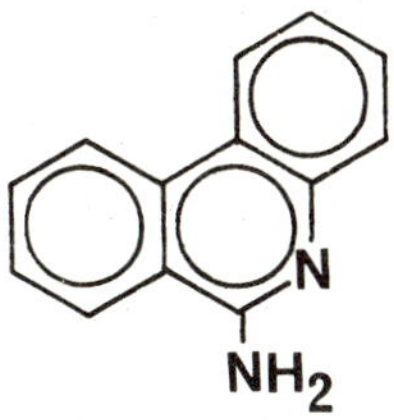

9-Aminophenanthridine

It has been proposed that <u>peri</u> interactions and an <u>ortho</u> steric hindrance, respectively, operate to produce these aberrations. When these two points are omitted from the plot the correlation coefficient is improved to a quite satisfactory value of 0.959. For this plot the slope, $(h_{N^+} - h_N)$ $\beta/2.303$ RT is 17.57. Given the usual thermochemical value for β (18-20 k cal/mole) one calculates that $(h_{N^+} - h_N)$ is in the range of 1.2 to 1.3.

The Union of π-Systems

The secular determinant for classical butadiene (p. 23) differs from that for the delocalized system (p. 21) in the absence of the off-diagonal elements for the resonance integrals between carbons 2-3. As far as classical butadiene is concerned, the π-system is equivalent to two isolated ethylenes. The union of two ethylenes to form delocalized butadiene represents a perturbation in which β_{23} goes from zero to some finite value.

Just as in our earlier examination of first-order perturbation theory, systems with degenerate energy levels cause some difficulties and our initial considerations will be directed to systems without such degeneracies, i.e., we will avoid the case of the union of two ethylenes for the moment. Consider two isolated π-systems R and S which undergo union to form a delocalized system R-S. Let us define the following:

$$H^{RS} = H^{R} + H^{S} + H^{1}$$

where H^{RS} is for the delocalized whole composed of H^{R} for R, H^{S} for S, and H^{1} for the perturbation.

For R

$$\Psi_m = \sum_i a_{mi} \psi_i$$

$$E_m = \int \Psi_m H^{R} \Psi_m \, d\tau$$

For S

$$\Phi_n = \sum_j b_{nj} \phi_j$$

$$F_n = \int \phi_j H^{S} \phi_j \, d\tau$$

By the orthogonality-normality conditions

$$\int \Psi_m \Psi_n \, d\tau = \delta_{mn}$$

$$\int \Phi_m \Phi_n \, d\tau = \delta_{mn}$$

$$\int \Psi_m \Phi_n \, d\tau = 0$$

Also

$$\int \Psi_m H^1 \Psi_n \, d\tau = \int \Phi_m H^1 \Phi_n \, d\tau$$

and

$$\int \Psi_m H^1 \Phi_n \, d\tau = a_{mr} b_{ns} \beta_{rs}$$

The latter integral comes about because the union occurs only between atoms r and s, and the new bond is characterized with β_{rs}.

Subject to the proviso of no degeneracies the perturbations to E_m and F_n are given by first-order perturbation theory as

$$\delta E_m = \int \Psi_m H^1 \Psi_m \, d\tau = 0 \qquad (5\text{-}24)$$

and

$$\delta F_n = \int \Phi_n H^1 \Phi_n \, d\tau = 0 \qquad (5\text{-}25)$$

While we have not considered second-order perturbation theory, extension of our previous considerations to the λ^2 terms leads to a second-order correction to the energy of the form

$$\delta E_m^{(2)} = \sum_n \frac{a_{mr}^2 b_{ns}^2 \beta_{rs}^2}{E_n - F_n}$$

$$\delta F_n^{(2)} = \sum_m \frac{a_{mr}^2 b_{ns}^2 \beta_{rs}^2}{F_n - E_m}$$

In view of the fact that the bonding MOs of AHCs do not vary greatly in energy, it has been suggested that this second-order correction might be simplified by using some mean value $\overline{E}$ for energy difference terms leading to

$$\delta E^{(2)} = E_{R+S} + \frac{\beta_{rs}^2}{2\overline{E}} \qquad (5\text{-}26)$$

While the first-order result above may appear disappointing, in the subsequent chapter we will see that the result appears to correspond to reality.

We can demonstrate the first-order approximation by considering the union of ethylene with butadiene to form benzene. Since there will be no perturbational correction, the energy of the π-system will simply be the sum of the energies of the parent systems, i.e., $2(\alpha + \beta)$ from the ethylene and $2(\alpha + 1.618\ \beta) + 2(\alpha + 0.618\ \beta)$ from the butadiene for a total of $(6\ \alpha + 6.472\ \beta)$. This figure is to be compared with the HMO value of $(6\ \alpha + 8\ \beta)$.

When the stricture regarding degeneracy is dropped the first-order change in energy ceases to be zero. The case of interest regards the union of two odd AHC fragments. The change in π-energy in this case being

$$\delta E = 2a_{or}b_{os}\beta = 2p_{rs}\beta \qquad (5\text{-}27)$$

where a_{or} and b_{os} are the NBMO coefficients of atom r in R and atom s in S.

Equation (5-27) may be illustrated by the union of methyl with benzyl to form styrene.

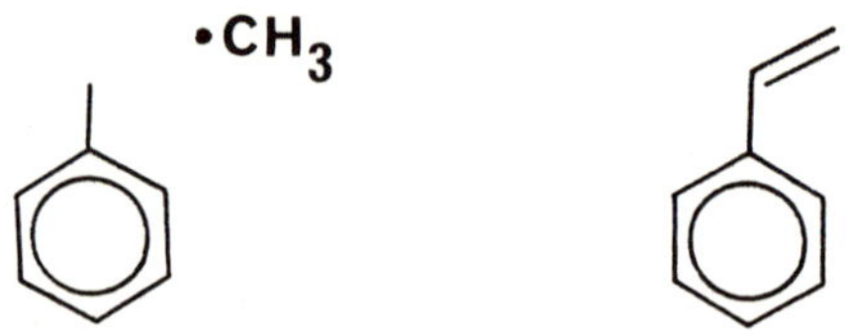

$$\delta E_{\pi} = 2\ (1 \times 2/\sqrt{7})\ \beta = 1.51\ \beta$$

In this case, the single carbon fragment has a coefficient 1 due to normalization, while the coefficient for the external carbon NBMO was derived on p. 42. The HMO value for E benzyl is 7 α + 8.72 β, and that for styrene is 8 α + 10.42 β. Thus, the HMO value for δE_{π} is 1.70 β

(the energy of the electron localized on the one carbon fragment being just α). Whether one should consider this good agreement is a matter of opinion. It is certainly evident that the calculations on p.41 and Eq. (5-27) are much easier than finding the roots for a 7 x 7 and a 8 x 8 pair of determinants.

The above treatment of PMO theory is, at best, an introduction. Dewar (1952) and others have greatly extended the method, and in the chapters to come we shall state many of these results without going into the detailed derivations.

Exercises

1. According to PMO theory the total π-energy on union of two HCs is

$$E_{\pi} = E_R + E_S + E^1$$

where E^1 is the perturbation energy. For AHCs the value of E^1 was shown to be zero. Calculate E_{π} for butadiene assuming (1) that it is formed by the union of two ethylenes and (2) that it is formed by the union of methyl with allyl. Compare these results with the HMO value.

2. Hexatriene can be considered to be formed by the union of two allyls or by the union of methyl with pentadienyl. Calculate E_π by each path.

Chapter 6

AROMATICITY, ANTIAROMATICITY, AND RESONANCE

General Comments

It has been well over a hundred years since Kekule, struggling to reconcile the molecular formula for benzene with its peculiar chemical stability, suggested the ring structure with alternating single and double bonds. It was soon recognized that benzene, naphthalene, anthracene and related structures all possessed the common traits of stability to oxidation, reluctance to undergo hydrogenation, diffidence to enter into halogen addition reactions, but a willingness toward electrophilic substitution.

In the intervening years, chemists have played the game of attempting to use one set of characteristics or another to define this concept of aromaticity or aromatic character. A variety of physical properties have been suggested at one time or another. Among these may be listed the absorption of light at wavelengths longer than those found for olefins, the anisotropy of magnetic susceptibility of molecular crystals and the downfield position of the NMR chemical shifts of protons attached to the carbon ring system. None of these have proved totally satisfactory.

As the concept of the covalent electron bond matured during the early part of this century, Robinson introduced the concept of the "aromatic sextet," i.e., six unsaturation electrons in a ring, as a structural feature of aromatic systems. Pauling and Wheland in the 1930s verbalized the constructs of valence bond theory to produce the pictorial resonance theory which still appears in most sophomore organic chemistry textbooks. According to this approach conjugated molecules are to be represented as the superposition of structures written according to the rules of valence.

The above device represented a step forward for the average chemist at work at the laboratory bench. He could "understand" why the bond lengths in butadiene were different from unconjugated dienes, why 1,4-additions occurred, and why the heat of hydrogenation of butadiene was less than expected on the basis of ethylene.

The development of HMO theory provided a more quantitative picture for unsaturated systems. The problem is that simple HMO calculations uniformly exaggerate the extent of electron delocalization and delocalization energies in conjugated systems, whether linear or cyclic. Because of this there is a failure in the usual molecular additivity properties such as bond lengths, bond energies, bond moments, heats of atomization, etc.

Experimentally, resonance or delocalization energies are calculated by comparison of heats of combustion or heats of hydrogenation between the conjugated system and some standard "localized" system, usually ethylene or cyclohexene.

During the course of certain self consistent field (SCF) calculations of a type considerably more sophisticated than the HMO variety, Dewar and his co-workers realized that, in fact, the collective properties of conjugated polyenes could be expressed as additive functions of newly defined quantities characterizing the carbon double and single bonds in these structures.

In contrast to benzene where two classical "Kekule" structures may be written, butadiene has only one classical structure:

$$CH_2 = CH - CH = CH_2$$

The bonds here are referred to as essential double and single bonds. According to HMO calculations butadiene and longer chain polyenes have relatively high DEs, and furthermore, the bonds near the center of such molecules tend to assume uniform intermediate length and bond orders. The latter observation is not found in the SCF calculations, the double and single bond pattern continue to alternate down the chain. Thus, if one has a characteristic value for the essential double and single bonds the thermodynamic properties of the polyenes may be calculated quite well.

It is the contention of Dewar that the flaw lies in our concept of the localized bond. Initially conceived to explain the directedness of the covalent bond and the geometry of molecules, the localized bond is usually considered to exist between only two atoms and to be composed of the overlap of atomic orbitals centered on those atoms. In point of fact, molecular orbitals always extend over the whole molecular framework. In the case of butadiene there is an interaction between the π-electrons associated with the two double bonds, but the same type of interaction occurs in all conjugated polyenes. One can, therefore, find collective properties for all of the conjugated polyenes which should still be considered to consist of an alternating pattern of essential double and single bonds. Dewar maintains that any hydrocarbon whose collective properties follow an additivity principle can be said to contain localized bonds. Butadiene and other polyenes for which only one classical structure can be written fall in this category though benzene, naphthalene, etc., do not.

Recently, Hess and Schaad have shown that this definition of the localized bond may be derived even from HMO calculations with an accuracy comparable to those achieved by the SCF results. They have derived a table of π-bond energies as follows (p. 80).
On this basis the energy for butadiene would be

$$E_{\pi} = 2\ E_{23} + E_{12} = 4.4660\ \beta$$

The HMO value (p. 20) is 4.48 β so that the DE is only 0.01 β at best. For benzene the calculation

Bond designation	Bond type	π-Bond energy (β)
23	$CH_2 = CH-$	2.0000
22	$-CH = CH-$	2.0699
22	$CH_2 = C<$	2.0000
21	$-CH = C<$	2.1083
20	$>C = C<$	2.1716
12	$>CH-CH<$	0.4660
11	$>CH-C<$	0.4362
10	$>C-C<$	0.4358

is

$$E = 3\,E_{22} + 3\,E_{12} = 7.608\,\beta$$

This is some 0.39 β below the HMO value. This new DE for benzene is smaller than older values because the new reference polyene - cyclohexatriene - has a larger binding energy.

The PMO Method, Localized Bonds, and Aromaticity

In the preceding chapter a brief introduction to the PMO method was given although many of the details were omitted in the interest of getting on to the results. It was shown, however, that the union of two even-AHCs to form RS resulted in no first-order change in the π-energy. As mentioned, the second-order change is $\beta^2/2\,\overline{E}$, where $\overline{E}$ was a mean value of the AHCs' bonding MOs.

The new bond between atoms r and s must be an essential single bond, for if it were a double bond - remembering that two even-AHCs are undergoing union - the remaining atoms in R and S would be odd in number. This line of argument goes along with our preceding section to the effect that bonds in classical polyenes are localized.

As we have seen, the first-order term is not zero for the union of two odd-AHC radicals. The same is true for the union between two atoms r and s in the same conjugated system. Here too, the energy change is given by

$$\delta E_{\pi} = 2\, p_{rs}\, \beta$$

Now one of the accepted theorems of the PMO method is that the bond orders between atoms of like parity in an even-AHC are zero. Parity in this context means that the atoms under consideration are either all starred or all unstarred.

An interesting application of these principles is the formation of fulvene, a non-AHC system, by either of the two processes shown here.

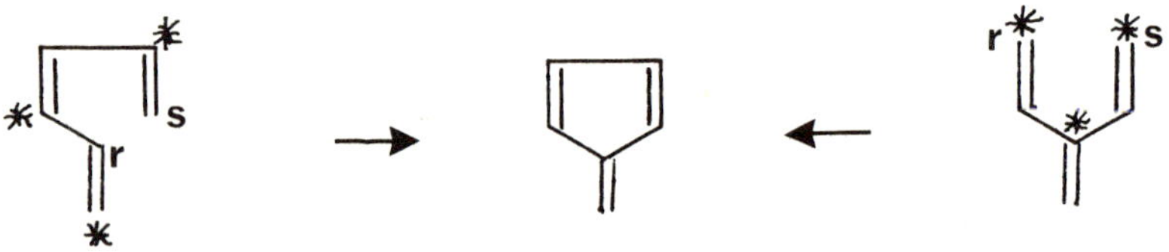

In either case the first-order change in π-energy will be zero. Following the same line of thought given for butadiene, either of the newly formed bonds will be essential single bonds. Fulvene is shown by the PMO method to be a cyclic polyene with localized single and double bonds. The HMO DE for fulvene was calculated in problem 2, Chapter 3. The DE derived from the localized bond approach of Hess and Schaad is given as an exercise to follow. Suffice it to say here, the chemistry of fulvene is that of a highly unstable polyene in agreement with the PMO and localized bond results.

The extension of the PMO method to the determination of aromaticity requires the expansion of Eq. (5-27) to the case of multiple union, i.e.,

$$E = 2 \sum_{rs} a_{or} b_{os} \beta = 2 \sum p_{rs} \beta \qquad (6\text{-}1)$$

A given cyclic system will be said to be aromatic if its π-energy is less than that of the corresponding linear polyene.

Thus, we may consider the union of two allyls to give benzene and hexatriene:

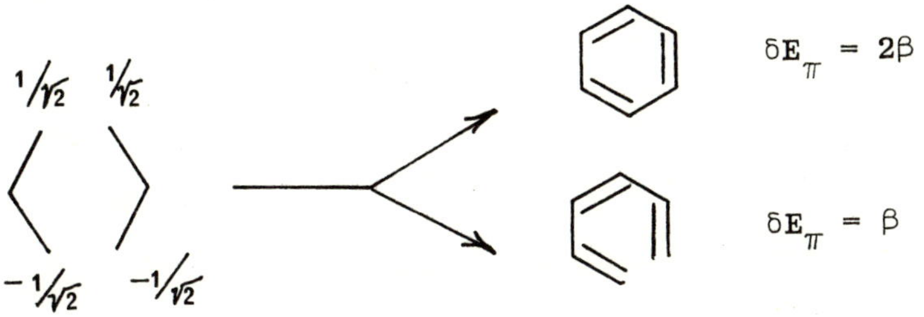

Since β is negative, benzene is the more stable of the pair.

A particularly simple form of (6-1) results if one fragment in the union is methyl:

$$E = 2 \sum_{s} a_{or} \beta \qquad (6\text{-}2)$$

The union of methyl with pentadienyl leads to the same conclusion regarding stability as that above.

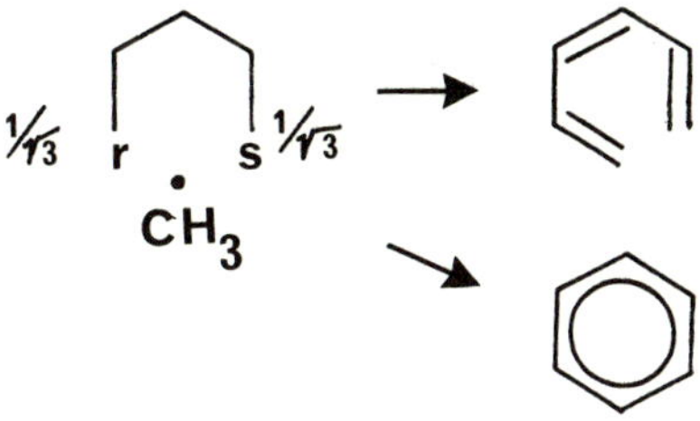

$$\delta E_\pi = 2\ (1/\sqrt{3})\ \beta = 2/\sqrt{3}\ \beta$$

$$\delta E_\pi = 2\ (1/\sqrt{3} + 1/\sqrt{3})\ \beta = 4/\sqrt{3}\ \beta$$

On the surface it might appear that a more direct assessment of the delocalization energy of benzene could be obtained by considering the union of the terminal carbons in hexatriene. These two carbons are not of the same parity and a finite bond order between them does exist. However, the determination of that bond order would require the determination of the coefficients of the MOs by the HMO method since the PMO method does not apply to even-AHs. The virtue of simplicity in the PMO method would be lost; hence, this is not considered a satisfactory approach.

A similar approach applied to the union of allyl and methyl yields cyclobutadiene.

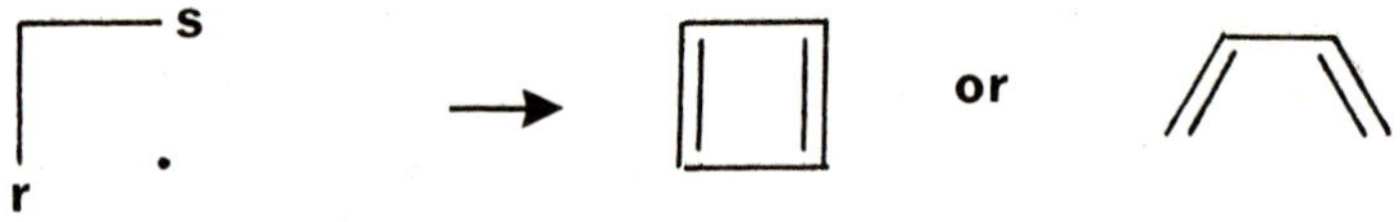

$$r = -s = 1/\sqrt{2}$$

$$\delta E_{\pi}^{\text{cyclobutadiene}} = 2\ (1/\sqrt{2} - 1/\sqrt{2})\ \beta = 0$$

$$\delta E_{\pi}^{\text{butadiene}} = 2\ (1/\sqrt{2})\ \beta = 2/\sqrt{2}\ \beta$$

Cyclobutadiene is less stable than its open chain analog which means it would actually have a negative resonance energy. Such compounds are said to be "antiaromatic." Many attempts over the years to prepare cyclobutadiene failed. However, it was finally prepared by Petit at the University of Texas and found to have a fleeting existence in the cold and in dilute solutions. Metal coordination complexes with cyclobutadiene as a ligand have been prepared.

The Hückel Rule

During the early 1930s Hückel produced an empirical rule from his considerations of the electronic structure of cyclic polyolefins. Those cyclics with (4n+2) π-electrons were predicted to show the properties of aromatic compounds, while those with 4n π-electrons would be nonaromatic. The n here is any integer, i.e., rings with 2, 6, 10, etc., electrons are to be aromatic. This stood in contrast to HMO results where all cyclic polyolefins (other than cyclobutadiene, DE = 0) were found to have appreciable DEs. Subsequent experimental work has supported the validity of the Hückel rule.

The PMO method can be shown to be in complete accord with the Hückel rule and, perhaps, can be said to provide, in theory, a real basis for its operation. Consider a linear conjugated radical of 2n - 1 carbons. Single union with methyl will yield the 2n polyene, while double union gives the cyclic 2n annulene. The NBMO coefficients for the linear odd-AHC vary in a systematic fashion, i.e.,

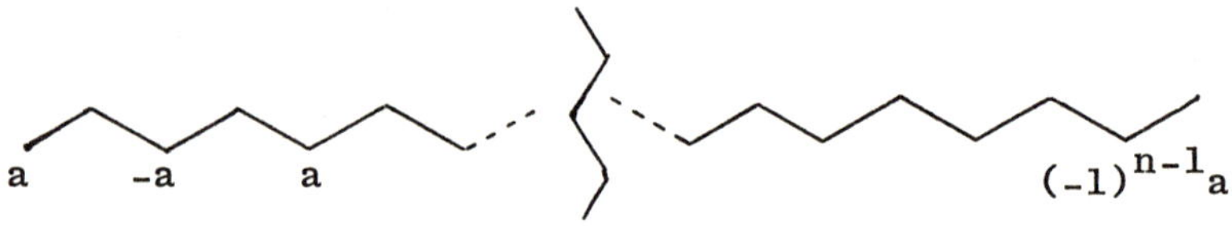

$$a = 1/\sqrt{n}$$

For the linear polyene

$$\delta E_{\pi} = 2\ a\ \beta$$

For the annulene

$$\delta E_{\pi} = 2\ (a + (-a)^{n-1})\ \beta \qquad (6\text{-}3)$$

Now for the cyclic system

$$DE = \frac{2(-1)^{n}\ \beta}{\sqrt{n}} \qquad (6\text{-}4)$$

When n is odd the annulene is more stable than the polyene (DE is positive), but when n is even the reverse is true and the system is antiaromatic.

In its initial form the Hückel rule was meant to apply only to monocyclic systems, though one often finds misapplications to polycyclic compounds. Using the PMO method it is possible to systematically apply the Hückel rule to a polycyclic compound a ring at a time. The argument, briefly, is as follows. If one denotes three contiguous atoms, r, s, and t in one ring of a polycyclic polyene then it can be shown that the signs of a_{or} and a_{ot} will be the same if the ring contains (4n+2) <u>atoms</u>. Thus, the union energy to atom s will be positive, and the ring aromatic. If the ring contains 4n atoms then the signs of the coefficients will be opposite. A 4n ring will be antiaromatic.

Biphenylene serves as an interesting and convenient system to illustrate these points. The HMO prediction for biphenylene is a larger DE (4.51 β) than for biphenyl (4.38 β).

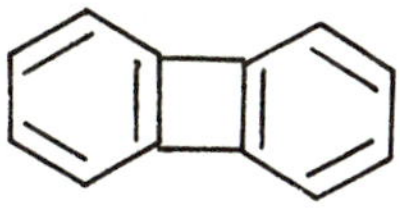

Biphenylene

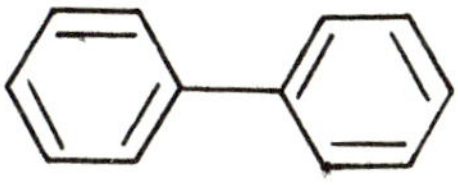

Biphenyl

According to PMO theory the union of two even-AHCs will, to first order, not produce a further reduction in E_π. The bond of union between the two rings should be an essential single bond and the DE is simply that of the two parts.

Biphenylene, biphenyl, and the open chain analog 1-phenyl-1, 3, 5-hexatriene may be constructed as follows:

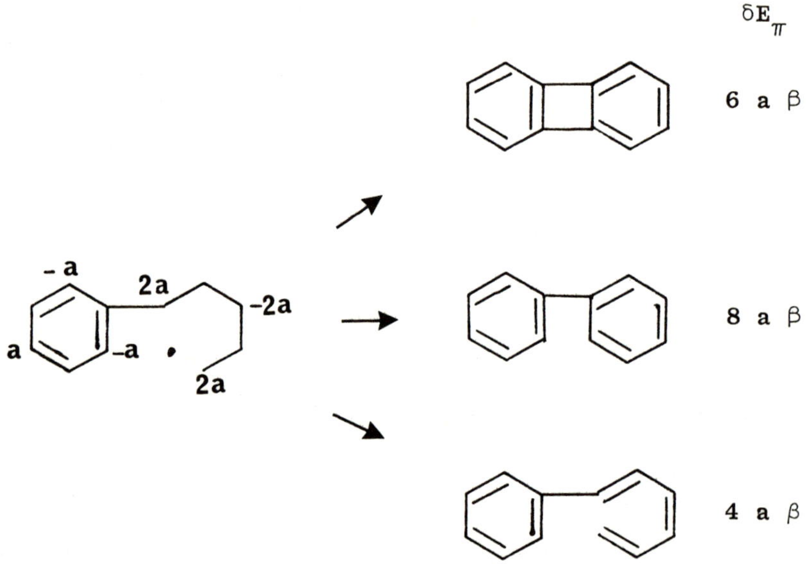

The π-energy of biphenylene is less than biphenyl. The two terminal six-membered rings are aromatic, while the central ring is antiaromatic. SCF calculations agree with the PMO result as opposed to the simple HMO result.

Nonalternant Hydrocarbons

Nonalternant hydrocarbons can be handled in much the same way as benzene when one has simple structures such as fulvene (p. 82). However, with more complex non-AHCs one must exercise certain cautions. Again it is convenient to subdivide the class into even or odd depending on whether or not there is an even number of conjugated atoms.

Azulene is an even non-AHC with two odd-numbered rings. An instructive series of compounds, including azulene, can be formed from nonatetraenyl and methyl according to the scheme below. The values shown are δE_{π}.

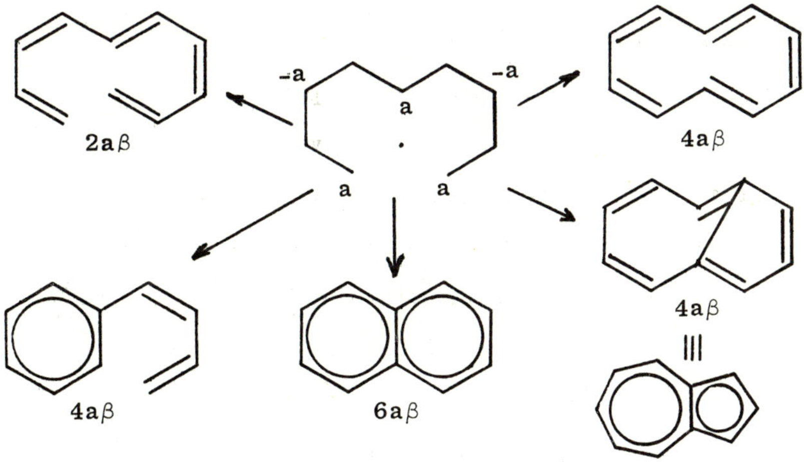

As can be seen, azulene has an appreciably lower DE than naphthalene. Furthermore, the DE is the same as the monocyclic polyene decapentaene, a (4n+2) π-system. The bond fusing the two rings must be an essential single bond, and bond length measurements indicate this to be so.

In order to deal with odd-AHCs it is necessary to extend the rules above. Thus, the theorem regarding bond orders between nonadjacent atoms in the same molecule (p. 81) must be expanded to cover the fact that bond orders in odd-AHC radicals are zero also for the union of atoms of like parity. The internal union of an allyl radical or pentadienyl radical to give the cyclopropenyl or cyclopentadienyl radicals respectively will result in no π-energy change since the bond orders between the two atoms uniting are zero.

When the cyclic odd-AHC is present as a cation or anion, however, there will be a first-order π-energy effect when compared with the open chain analog. It is this open chain analog which serves as the comparison compound in making statements about aromatic or antiaromatic character. With this understanding one may now write for the anion $\delta E_{\pi} = 2\ p_{rs}\ \beta = 2\ a_{or}\ a_{os}\ \beta$ and for the cation $\delta E_{\pi} = 2\ p_{rs}\ \beta = -2\ a_{or}\ a_{os}\ \beta$. As an example, we may use the pentadienyl system. Since β is negative, fusion to the anion results in a stabilizing change in π-energy.

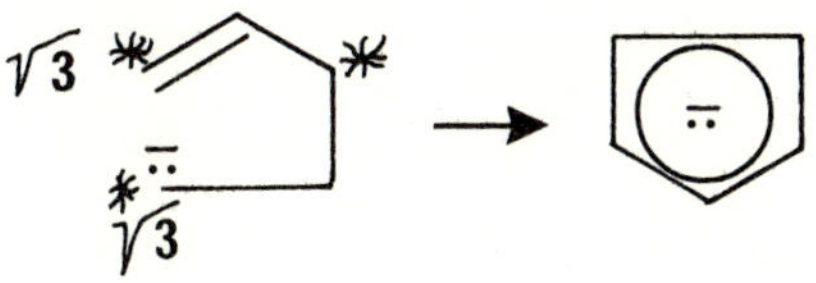

$$\delta E_{\pi} = 2(1/\sqrt{3} \cdot 1/\sqrt{3})\ \beta$$

$$\delta E_{\pi} = -2(1/\sqrt{3} \cdot 1/\sqrt{3})\ \beta$$

From the equations above, it is clear that systems where the coefficients have the same sign will be aromatic while those in which the signs are different will show the cation as being the aromatic form. The relative signs of the coefficients can be determined by inspection of the open chain form or from the cyclic product. If the new ring has (4n+1) atoms, the coefficients will be alike. Rings with (4n+3) atoms will have opposite signs for the open chain coefficients. So it follows that (4n+1) rings will have aromatic anions but antiaromatic cations, and the reverse will be true for (4n+3) systems. In either event, the aromatic system will have (4n+2) π-electrons.

Resonance Theory and MO-PMO Concepts

As implied earlier, the beauty of the theory of resonance, as conventionally taught, is in its graphical nature which allows the chemist to "think" on his feet. The fact that cyclobutadiene may be written with two Kekule structures or that relative stabilities for biphenyl and biphenylene are reversed from their structure counts of four and five are considered as aberrations that one must learn to live with. The best that one can do with resonance theory, in any event, is make qualitative statements.

Initially resonance theory was put forward as an extrapolation of the more quantitative valence bond theory. Since both involve the examination of structures written according to the rules of the electron pair bond this implication is still often given in texts on organic chemistry. Over the years, however, a growing literature has established the close conceptual connection between resonance theory and MO theory.

We have just gone through the argument that for polyolefins such as butadiene, where only one structure can be written, MO theory, whether in the SCF or PMO approximations, predicts very little stabilization due to electron delocalization. In contrast, benzene with its two Kekule forms shows physical and chemical properties intermediate to either and appreciable delocalization energy. These results can be generalized to other linear and cyclic polyenes.

It has been known for some two decades that many of the anomalies of resonance theory can be resolved by virtue of the fact that it is possible to attribute signs to the various resonance contributing forms. Forms with differing signs have opposite effects on stability. Thus, although one may draw more resonance forms for one molecule than for another, statements about the relative stabilities of the molecules are not warranted unless the signs of the various contributors are taken into account.

Consider first one of the Kekule structures for benzyl. Star the structure as an odd-AHC in such a way that the starred positions outnumber the unstarred. Number the unstarred positions in any order from 1 to n.

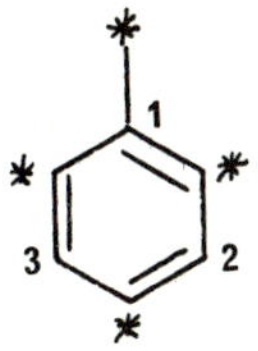

The starred set are numbered from n+1 to 2n+1. Now

form pairs of one unstarred and one starred atom by pairing the doubly bound atoms of the particular structure in question. For this purpose, the odd atom may be considered to be paired with a fictitious unstarred atom 0.

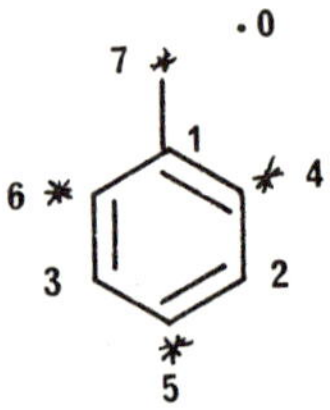

The bonding arrangement for the above example is represented by the set (0 7)(1 4)(2 5)(3 6). Now arrange the pairs in increasing order of the unstarred atoms 0 to n. By pairwise exchanges, rearrange the numbers n=1 to 2n+1 (the second number of each pair) in ascending sequence. The above structure is defined as even (plus) or odd (minus) according as an even or odd number of permutations of atoms is necessary.

(0 7)(1 4)(2 5)(3 6) ⟶ (0 4)(1 7)(2 5)(3 6) ⟶
(0 4)(1 5)(2 7)(3 6) ⟶ (0 4)(1 5)(2 6)(3 7)

Since three permutations are required, the sign of

the structure written above is defined as negative.

It is possible to show that the other possible Kekule structure is also negative. In comparing different structures it is important that the original numbering system be maintained. Similar treatments where the odd atom is in the ring lead to the following results.

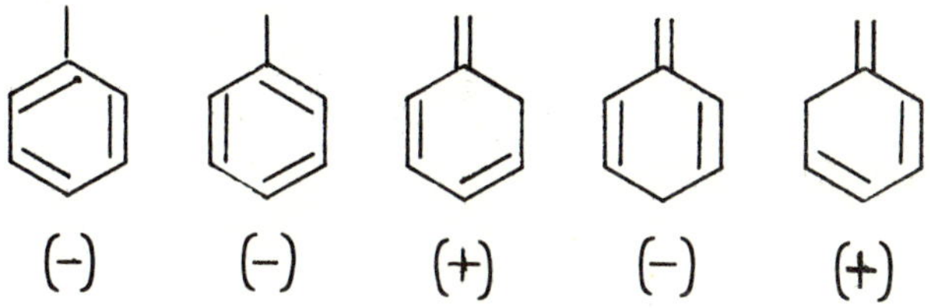

This specific case illustrates a more general point, namely, that structures which differ only in the orientation of one bond (such as 2 and 3 above) are of opposite sign.

Proceeding through an argument which we shall not reproduce, it is possible to show that for an odd-AHC the number of classical structures of the same sign with the odd atom in a given position is equal to unnormalized NMBO coefficient at that position.

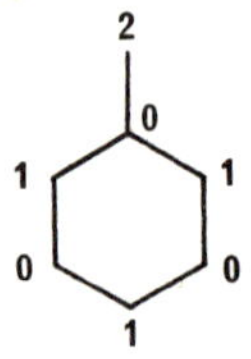

We know the odd electron in the benzyl radical has its highest probability density at those active sites where the amplitude of the NBMO coefficients is greatest. One sees here a direct link between resonance theory and MO results. Obviously, the same reasoning may be carried over to the case of the benzyl cation or anion.

Further generalizations for the odd AHC case may be cited. Consider pentadienyl and α-methylnaphthyl along with benzyl from above.

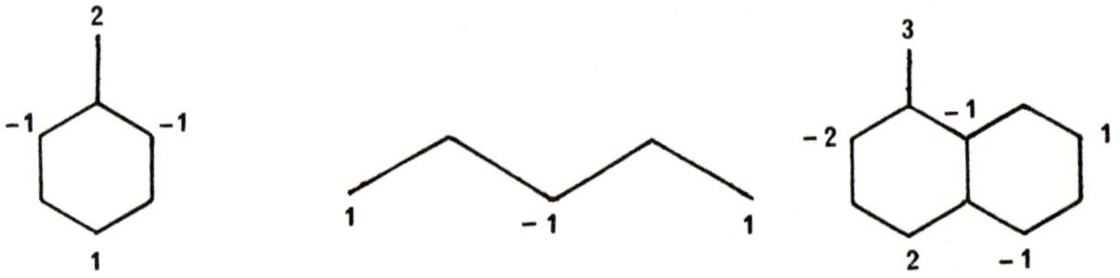

The total number of classical structures for each case is given by the sum of the absolute values of the coefficients. Their unions of methyl yield for example styrene, hexatriene, and α-vinylnaphthalene. These compounds have a total of two, one, and three classical structures respectively, just the value of the coefficient at the point of union. The same conclusion holds for the removal of the odd atom from the conjugated system; benzene, butadiene, and naphthalene also showing two, one, and three structures. These observations point the way to how we might handle resonance counting in even AHC systems.

The point of importance is not the absolute sign associated with a given group of structures, but rather whether the signs are the same or different throughout the group.

Let us consider cyclobutadiene. We could do it as above using methylenecyclobutadiene. Instead we will treat the two Kekule forms directly.

(13)(24) ⟶ (12)(34)

There is one permutation. Therefore, the structure A is opposite in sign from B.

In our earlier HMO treatment of cyclobutadiene (p. 34) two NBMOs were discovered in what was clearly an even-AHC. It can be shown that in even-AHCs where half the structures are of one sign and half the other similar NMBOs, equal in number to the structures written, will be found. These orbitals are termed "supernumerary" orbitals. Such molecules will exhibit diradical ground states if planar (cyclobutadiene) or will twist into a nonplanar puckered configuration (e.g., cyclooctatetrene). Benzenoid structures having (4n+2) membered rings will have classical structures all of the same sign and with no supernumerary orbitals.

The biphenylene case may be used to make another point. The resonance structures are given here.

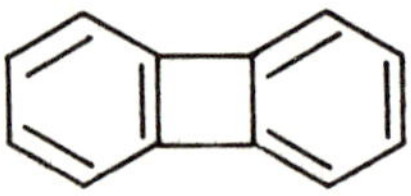

A

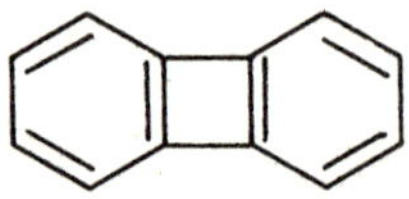

B

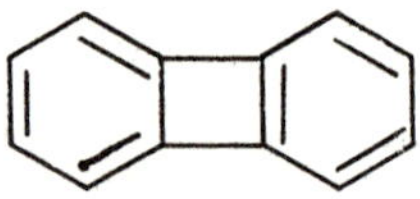

C

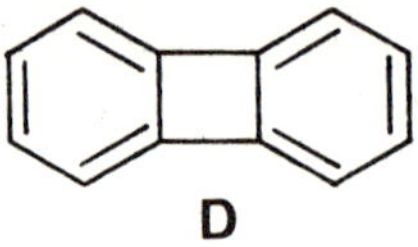

D

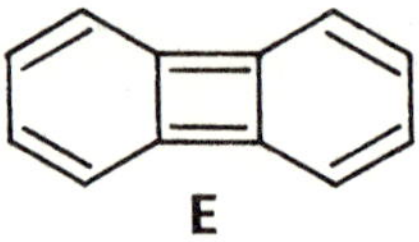

E

The signs of each structure could be determined as for benzyl by adding an α-methylene group, or the relative sign of each structure with reference to A may be ascertained as previously for cyclobutadiene. In either event, one concludes that A, B, C, and D are of one sign and E the opposite. The algebraic structure cound (ASC) is the difference between the number of structures of opposite sign. For biphenylene the ASC is 3. The structure count (SC) is five. In biphenyl all structures are of the same sign since both rings are benzenoid, thus the ASC and SC are the same. The resonance theory with signs correctly assigned predicts biphenyl (ASC 4) to be more stable than biphenylene (ASC 3). The result, of course, is in accord with PMO conclusions.

Another generalization may be offered to handle even-AHCs with one or more 4n rings provided they are not fused to each other. The ASC in such systems is that number remaining after all structures having 2n double bonds in the 4n ring(s) have been deleted. For instance, I has been found to be more stable than II.

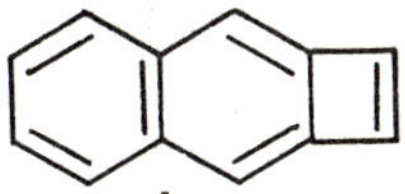

I

ASC 2

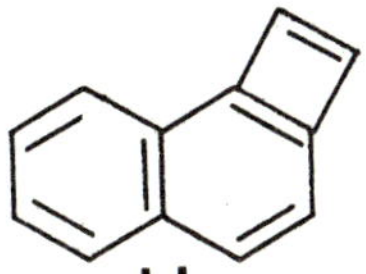

II

ASC 1

Yet another technique exists for determining values of SC and ASC. For even-AHCs one may delete an atom generating an odd system. With fused ring systems one must choose an atom at a fusion point. The NBMO coefficients for the odd-AHC are then determined. The sum of the absolute values of the unnormalized coefficients about the deletion point equals the SC; while the absolute value of the sum equals the ASC.

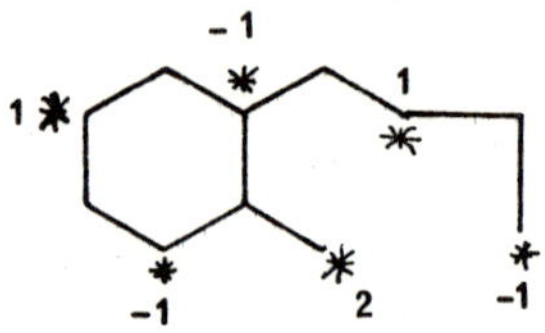

SC4, ASC 2

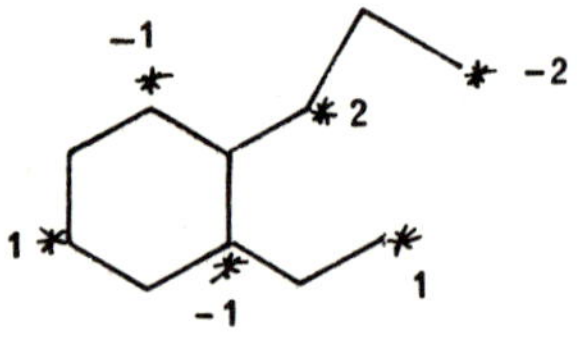

SC 5, ASC 1

Interestingly, the atom deletion technique works quite well with non-AHCs. Application to azulene (as well as for naphthalene for comparison purposes) is shown below.

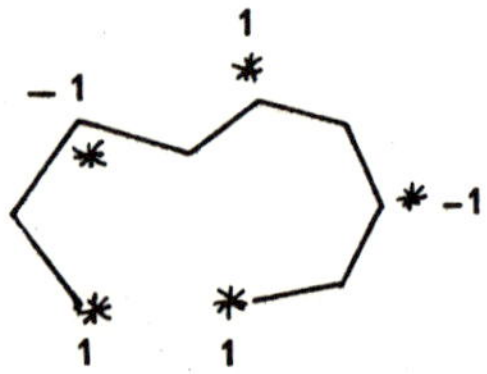

ASC 2

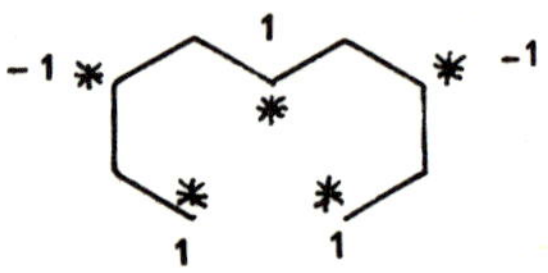

ASC 3

According to our earlier PMO treatment the multiple union of two odd-AHCs to form an even-AHC is accompanied by an energy change

$$\delta E_{\pi} = 2(a_{or} b_{os} + a_{oq} b_{ot}) \beta \quad (6\text{-}5)$$

where a_{or}, a_{oq} and b_{os}, b_{ot} are the appropriate NBMO coefficients in the two fragments. Now if the two fragments are benzenoid, the absolute values of the unnormalized coefficients expressed as whole numbers are the SCs with the odd atoms located at r and s in one and at q and t in the other. Equation (6-5) may be rewritten as

$$\delta E_{\pi} = 2K(n_r m_s + n_q m_t) \beta \quad (6\text{-}6)$$

where the ns and ms are the structure counts. The quantity in brackets in (6-6) is the total number of classical structures which can be written for the united molecule.

Using benzyl as an example one may form each of the following:

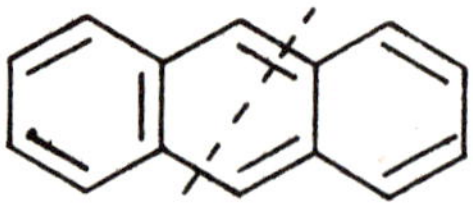

SC 4
(2·1 + 2·1)

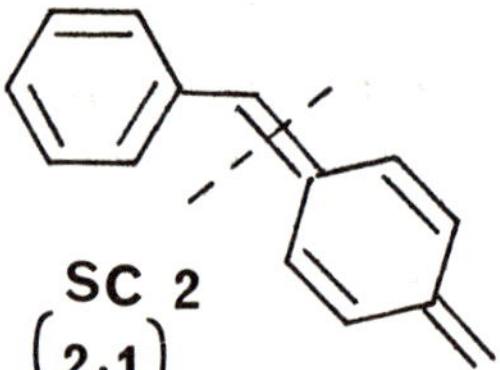

SC 2
(2·1)

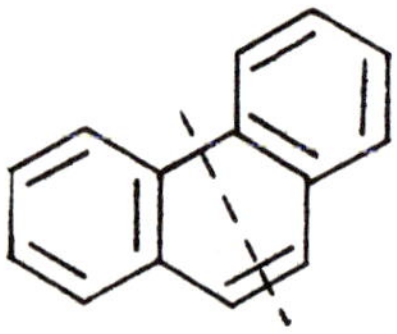

SC 5
(2·2 +1·1)

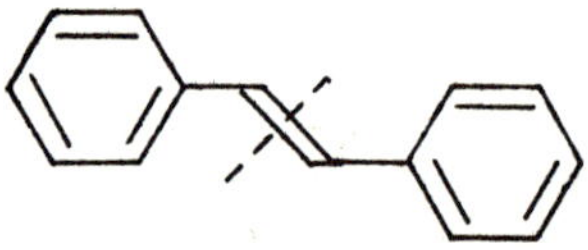

SC 4
(2·2)

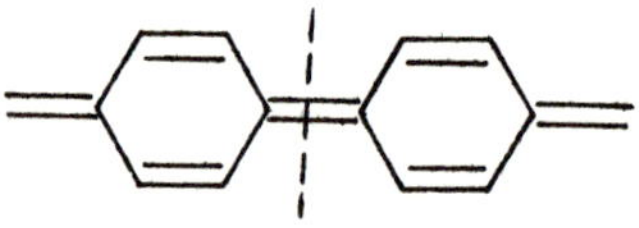

SC 1
(1·1)

The union energy for phenanthrene is greater than for anthracene, a result in keeping with experimental resonance energies and the generalization that angular fused systems are more stable than their linear counterparts. The relative stabilities of the quinoid structure are also as expected.

Exercises

1. Using the bond energies of Hess and Schaad calculate E_π for localized fulvene. From the HMO value of E_π (Chapter 4, problem 2) calculate the

DE for fulvene. Contrast this with the conventional DE calculated in the earlier problem.

2. Using the PMO method calculate the DE for fulvene by the union of methyl with pentadienyl. [Note: the comparison compound in this case is hexatriene.]

3. Calculate the DEs for biphenyl, biphenylene, naphthalene, and axulene using the localized bond energies of Hess and Schaad. The HMO values for E_π are 16.38, 16.51, 13.68, and 13.36 β respectively. Compare these DEs with those predicted on the usual Hückel basis. Also compare the results with PMO predictions.

4. Calculate σE_π for octatetraene, cyclooctatetraene, styrene, benzocyclobutadiene (below), and pentalene (below) and compare the relative stabilities of these molecules. Use the union of heptatrienyl and methyl for one series of calculations. Then repeat using the union of pentadienyl and allyl. [Warning: Application of the PMO method requires the application of common sense as well as the rules. Bonds will tend to

form between active positions in preference to inactive positions.]

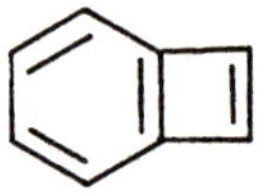

Benzocyclobutadiene

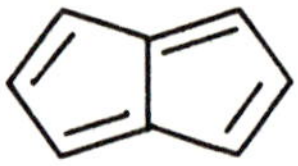

Pentalene

5. Determine the SC and the ASC for each of the molecules below. Are the results consistent with the fact that only IV and V are known molecules?

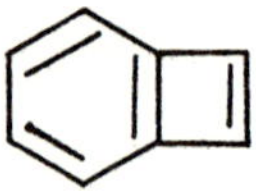

I

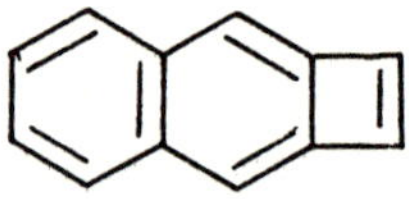

II

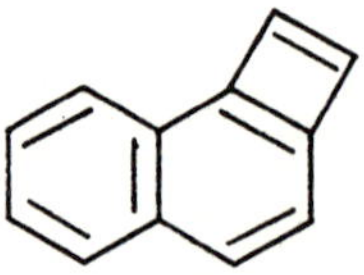

III

IV

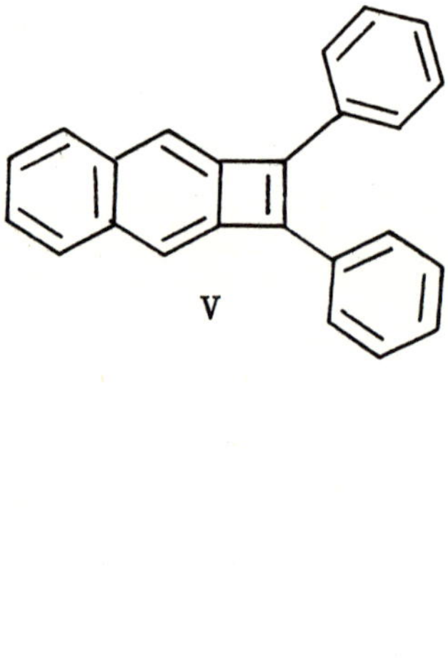

V

Chapter 7
CHEMICAL REACTIVITY

The Nature of the Problem

The prediction of reaction products, rates of reaction, and reaction equilibria is one of the prime goals of quantum chemistry. However, at present it is not possible to calculate reaction rates from first principles except for perhaps the simplest molecular systems. What is required, of course, is a knowledge of the free energy of reaction for equilibria and the free energy of activation for rate processes. The latter requires a knowledge of the transition state which, in turn, implies information of the whole potential energy surface for the reaction.

The breakdown of free energy into enthalpy and entropy offers no help directly in the solution of the problem since entropy calculations require a knowledge of vibrational and rotational frequencies for reactants and transition states which again are available only for very simple molecules.

More progress is possible by asking questions about relative reactivities. How reactive are various positions within the same molecule towards a given reagent? What differences in reactivity occur among a series of related molecules undergoing the same reaction? Inherent in any answer to these questions is the assumption either that all entropy changes are the same in the systems under examination or that there is some systematic variation in entropy which is compensated for in the enthalpy term. Experimentally it has been demonstrated that entropies of reaction and activation entropies often vary widely among the members of a reaction series. Happily there are many cases known where linear plots of entropy vs enthalpy are found. Organic chemists have a surprising list of reactions demonstrating logical correlations between reaction rates or equilibria and variations in molecular structure.

Finally, if we are to calculate reaction rates we need to have energy differences between the ground state and the transition state for the reaction. This implies a knowledge of the structure of the transition state which is sometimes hard to come by. In 1955 George Hammond gave formal expression to a rule of thumb found in many works dealing with rate phenomena. Hammond's postulate says in essence that a transition state with a free energy close to that of a reactive intermediate will have structural features closely resembling the intermediate. Since we have information about the reactive intermediates in many reactions, it follows that we often have knowledge of the nature of the transition states too.

Electrophilic Substitution

Substitution reactions in aromatic systems involving attack by an electron seeking reagent have been exhaustively studied over the years. Since much of the pertinent evidence regarding the mechanism can be found in organic chemistry texts, no review of the facts will be given here.

It is considered that attack on the π system by the electrophile Y^+ leads to the formation of a reactive intermediate with a pentadienyl cation structure. Collapse back to a fully aromatic product follows.

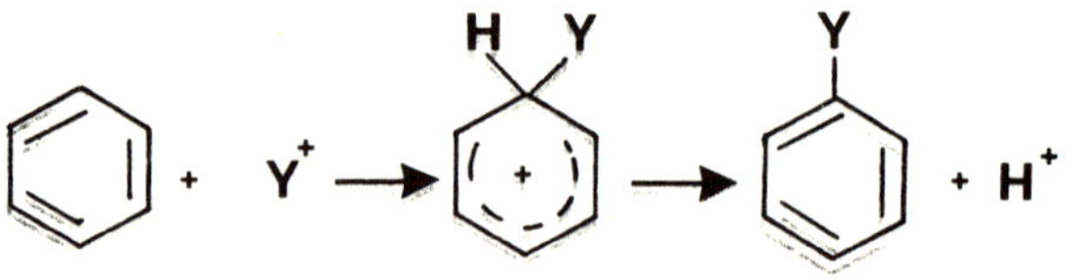

By Hammond's postulate the transition state for forming the high energy intermediate should look very much like the intermediate. If so, then the main energy change in reaching this transition state will be that of localizing two electrons from the π-system in the developing bond to Y. The loss of delocalization energy in going to the pentadienyl cation will measure this.

Wheland, some years ago, suggested the use of the localization energy as a useful reactivity index.

$$\Delta E_{loc} = E_{OA} - E_{EA} \qquad (7\text{-}1)$$

Here E_{OA} is the π-energy for an odd alternant system such as pentadienyl, while E_{EA} is the π-energy for an even alternant system of one more atom, such as benzene. From HMO calculations these values are $4\alpha + 5.46\ \beta$ and $6\alpha + 8.00\ \beta$, respectively. The value of the localization energy is 2.54 in units of $(-\beta)$ to keep the value positive. The α terms cancel since each localized p-electron contributes α to the energy sum.

The above calculation for substitution in benzene requires the complete solution of two secular determinants. The PMO method offers a far quicker solution. We saw earlier (p. 83) that the energy change in going from pentadienyl to benzene could be approximated by the union with methyl. This is precisely the energy change we require but with the sign reversed. Thus,

$$\Delta E_{loc} = -2\ \Sigma\ p_r\ \beta = -2\ \Sigma\ C_{or}\ \beta \qquad (7\text{-}2)$$

where the C_{or} are the NBMO coefficients of the positions being joined to the one carbon fragment. Applied to benzene:

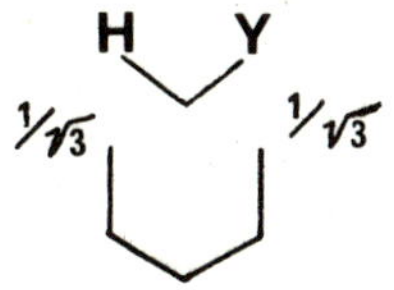

$$\Delta E_{loc} = -2\ (1/\sqrt{3} + 1/\sqrt{3})\ \beta$$

$$= -2.31\ \beta$$

Dewar has defined the term $2\ \Sigma\ C_{or}$ as N_t, the reactivity number. In general it may be said that values of N_t are smaller than HMO localization energies, but they parallel the latter and reflect reactivities as well.

Electrophilic attack on naphthalene can occur at either the α or β position.

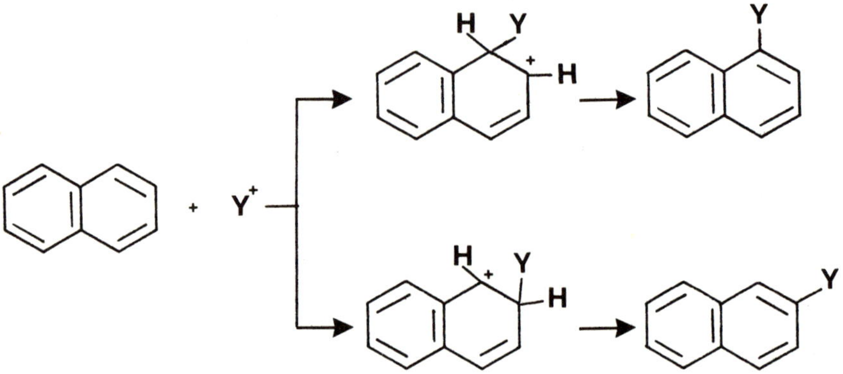

The energies of the two transition states may be approximated by the two systems below.

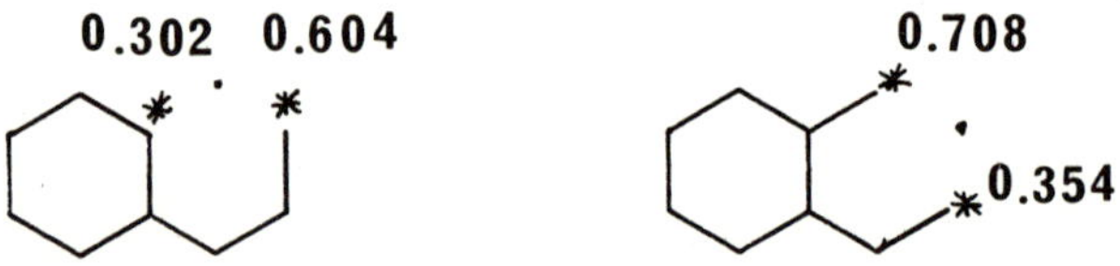

The numbers are the NBMO coefficients. The N_t for α substitution is 2(0.302 + 0.604) = 1.81, while that for β-substitution is 2(0.708 + 0.354) = 2.12. The lower reactivity number for α-substitution is in agreement with the experimental observation that α-products are kinetically first formed from naphthalene. For reversible reactions such as sulfonation one may actually isolate the thermodynamically more stable β-isomer.

The reactivity numbers for benzene, naphthalene, and anthracene are given below. Not only are the correct positions of substitution predicted for the latter two, but the relative ease of substitution in the series is correct.

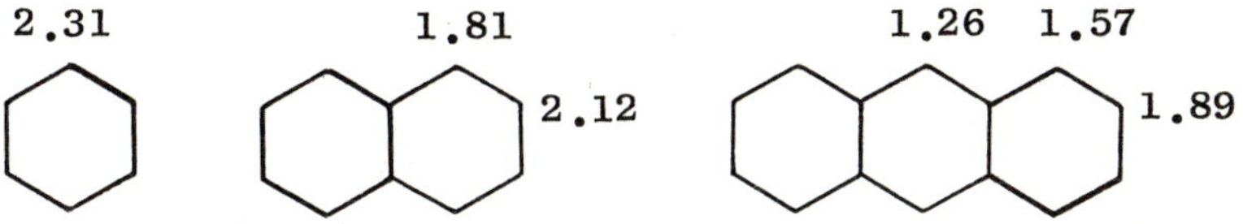

As a case in point Fig. (7-1) below shows the plot of the log of partial rate factors for nitration for a series of aromatic hydrocarbons against N_t.

Heteroatom Systems

As demonstrated in Chapter 5, heterocyclic molecules may be considered as perturbed versions of their carbocyclic analogs. In the case of reaction rate calculations we must know not only how the heteroatom perturbs the reactant but also

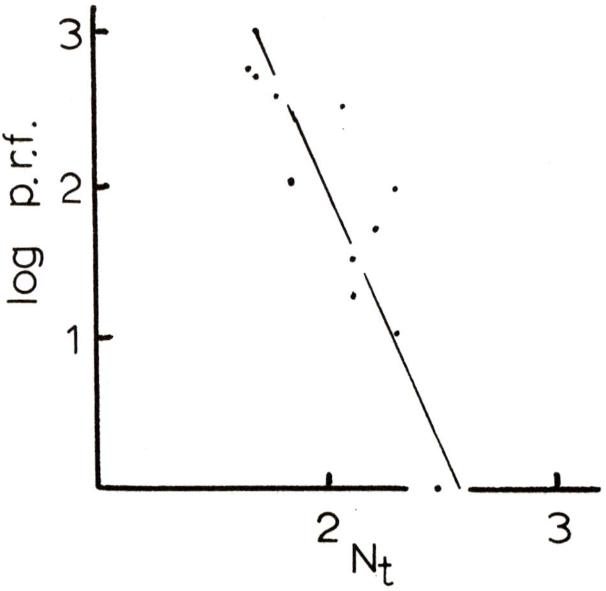

FIG. 7-1. Plot of partial rate factors (p.r.f.) vs the reactivity numbers (N_t) for a series of polynuclear aromatic hydrocarbons undergoing nitration in acetic anhydride.

how it perturbs the transition state. Let the change in π-energy on introducing the heteroatom into the even-AHC be δE_{EA}.

For the intermediate (an odd-AHC) let the same process be given by δE_{OA}.

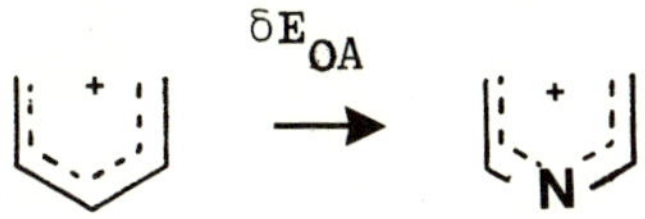

The localization energy for the electrophilic substitutions energy for pyridine may now be approximated by

$$\Delta E_{loc} = E_{OA} - E_{EA} + \delta E_{OA} - \delta E_{EA} \tag{7-3}$$

If we maintain the assumption that $\beta_{CN} = \beta$ from p. 63 and remembering Eq. (5-22), we can write the energy change for any orbital (u) upon introducing a heteroatom as

$$\delta E_u = C_{ui}^2 \; \delta \alpha_i \tag{7-4}$$

where again C_{ui} is the u-th MO coefficient of the atom i. The total change of energy (δE) requires multiplying δE_u by the number of electrons (N) in the orbital and then summing over all occupied orbitals:

$$\delta E = \sum_{u}^{occ} N_u C_{ui}^2 \delta \alpha_u = ED_i \delta \alpha_i \tag{7-5}$$

The term $\sum_{i}^{occ} N_u C_{ui}^2$ is just our earlier definition of total electron density. Since all carbons in a neutral AHC have electron densities of unity, Eq. (7-5) reduces to

$$\delta E_{EA} = \delta \alpha_i \tag{7-6}$$

for the conversion of benzene to pyridine.

The odd-AHC fragment above bears a charge. Consequently, the change in π-energy is dependent upon orbital occupancy. For odd-AHCs only radicals have unit electron densities at each position. A cation has one fewer electron than the radical, that electron being removed from the singly occupied NBMO. The electron density in the cation is then simply

$$2 \sum_{u=1} C_{ui}^2 = 1 - C_{oi}^2 \tag{7-7}$$

and therefore

$$\delta E_{OA} = (1 - C_{oi}^2) \delta \alpha_i \tag{7-8}$$

By virtue of (7-8) we can now get to the δE_{OA} term of Eq. (7-3) with only a knowledge of the NBMO coefficients.

Substituting these results into Eq. (7-3) allows us now to write for electrophilic substitution in the heterocycle

$$\Delta E^{+}_{loc} = \Delta E_{loc} - C^{2}_{oi}\,\delta\,\alpha_{i} \tag{7-9}$$

$$= -N_{t} - C^{2}_{oi}\,\delta\,\alpha_{i}$$

where ΔE_{loc} is the localization energy for the carbocyclic analog.

It may be well to mention here before proceeding that similar approaches can be taken to free radical and nucleophilic substitution reactions. The results for these cases have been shown to be

$$\Delta E^{rad}_{loc} = \Delta E_{loc} \quad \text{(free radical)} \tag{7-10}$$

$$\Delta E^{-}_{loc} = \Delta E_{loc} + C^{2}_{oi}\,\delta\,\alpha_{i} \quad \text{(nucleophilic)} \tag{7-11}$$

Let us now apply Eq. (7-9) to pyridine. The three possible intermediates representing their respective transition states are given below as are the NBMO coefficients for pentadienyl.

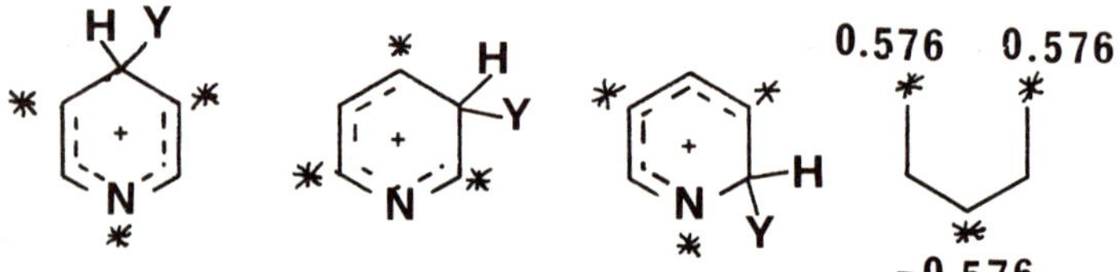

I II III

Note that in I and III the heteroatom appears at an active position in the pentadienyl fragment. Thus, the energies of these will be perturbed by $C^2_{oi}\delta\alpha_i$. Since for nitrogen this term is usually approximated by 0.5 β, we may write for each of the three cases above

$$\text{(I)} \quad \Delta E^+_{loc} = -2.31\ \beta - (0.576)^2 \times 0.5\ \beta$$

$$= -2.78\ \beta$$

$$\text{(II)} \quad \Delta E^+_{loc} = -2.31\ \beta - (0)^2 \times 0.5\ \beta$$

$$= -2.31\ \beta$$

$$\text{(III)} \quad \Delta E^+_{loc} = -2.31\ \beta - (0.576)^2 \times 0.5\ \beta$$

$$= -2.78\ \beta$$

The lowest value for the localization energy for electrophilic attack in pyridine is to be found in II, and, of course, this is where substitution does occur. However, the localization energy calculated here is the same as that for benzene itself and this stands in marked contrast to the experimental fact pyridine is considerably less reactive than the carbocycle. It has been speculated that the electronegativity of the nitrogen causes a general lowering of the electron density of the ring system compared to benzene and is reflected in a lowered reactivity at the 3-position. Another possibility is that of repulsion between a positively charged electrophile and the pyridinium ion which would be present in nitration or sulfonation reactions or other substitutions in acid media.

Free Radical Substitution

Detailed rate studies of free radical substitution processes are not available. However, relative reactivities towards methyl and trichloromethyl radicals have been ascertained. The best evidence supports a mechanism resembling that for electrophilic substitution. If so, then a similar relationship between reactivity numbers and the log of the reactivity index (called here the methyl affinity) should be noted. Since no product distributions have been determined for the polynuclear aromatic hydrocarbons, the plot shown in Fig. 7-2 uses the reactivity number at the most reactive position only.

Nucleophilic Substitution

Nucleophilic substitution reactions on substituted benzenes, in general, occur only with considerable difficulty. However, when a nitrogen

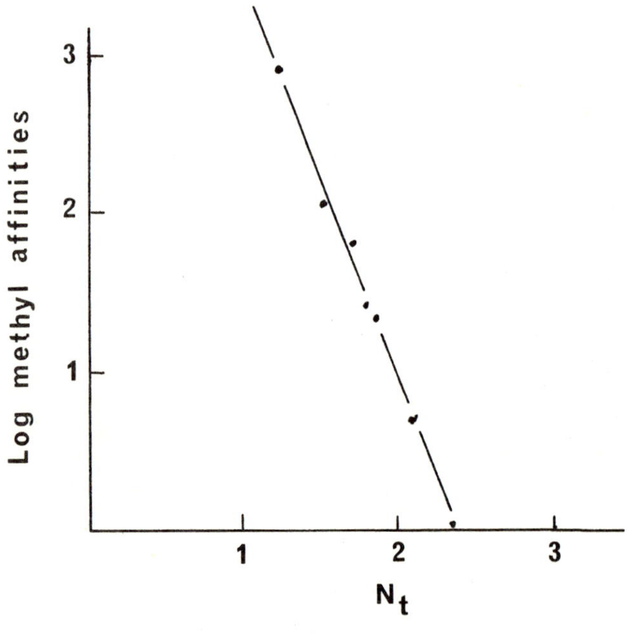

FIG. 7-2. Plot of reactivity numbers (N_t) vs the methyl affinities for a series polynuclear aromatic hydrocarbons.

atom is inserted in the ring such substitutions may take place readily.

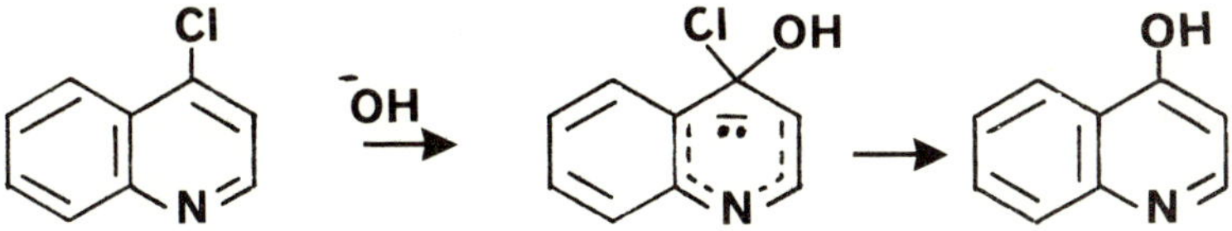

The situation is covered by (7-11), the delocalization energy being reduced by the $C_{oi}^2 \delta\alpha_i$ term. This term can only be important if the heteroatom is located at an active position in the intermediate. This, of course, means that the nitrogen must be ortho or para to the position of substitution. It goes without saying that molecules with multiple heteroatoms in strategic positions react even more readily.

Paralocalization Energies

In the next chapter we will consider in some detail a variety of multicenter reactions. These are characterized by the fact that several atomic centers in one or more molecules come together in a concerted breaking of old bonds and making of new ones.

One such reaction is the Diels-Alder reaction, and it is appropriate that we anticipate the subsequent discussion while considering localization energies.

The reaction of anthracene with maleic anhydride is an oft cited example of the Diels-Alder reaction.

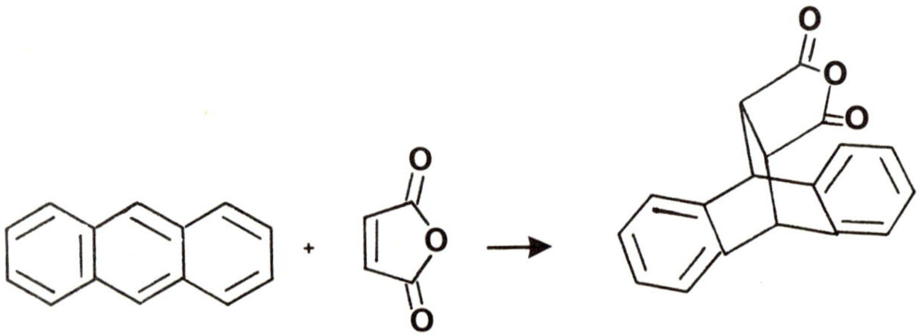

One may well wonder why the reaction occurs exclusively at the 9,10-positions.

Some years ago it was shown that such reactions correlate with the localization energies of the carbons para to each other in the aromatic systems. Paralocalization energies correspond approximately to the sum of the reactivity numbers for these positions (in units of $-\beta$). The values for anthracene are shown here, and not unexpectedly the paralocalization energy is lowest for the 9,10-positions.

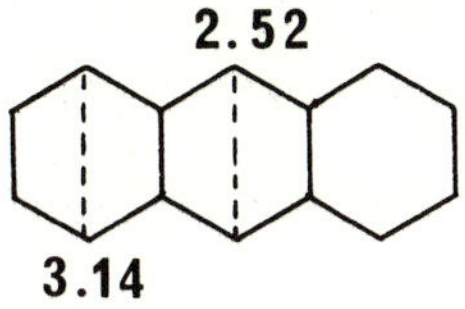

The paralocalization energies for naphthalene (3.62) and benzene (4.62) are considerably higher than the values for anthracene, and they do not form adducts under normal conditions.

<u>Ortholocalization Energy</u>

Certain aromatic hydrocarbons react with osmium tetraoxide to form osmate esters which can be hydrolyzed to diols.

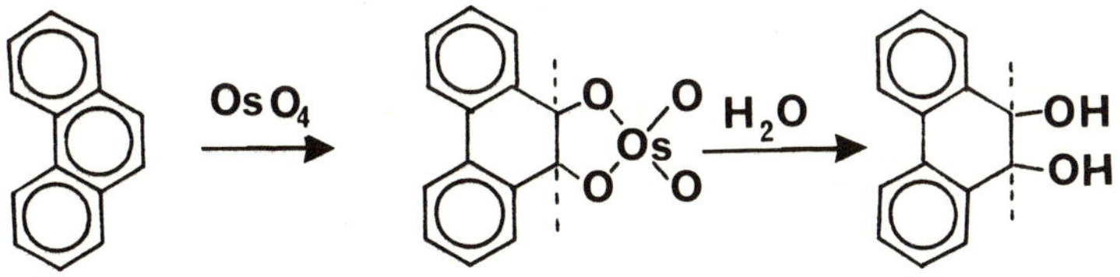

The mode of addition is <u>cis</u>, and clearly the reaction should be sensitive to the bond attacked as a portion of the conjugated system is destroyed. The effect on E_π of the loss of these two adjacent centers could be calculated from MO theory as in the case of the localization energy (7-1). However, Dewar has shown that a version of (7-2) can be used to obtain approximate ortholocalization energies, i.e.,

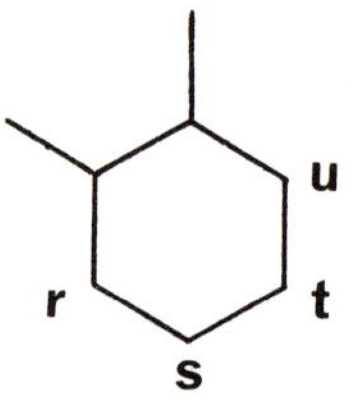

The requisite bond orders can be obtained, per (5-28), by union of the appropriate odd-AHC with methyl. As an example one might use phenanthrene. For the carbon 9 and 10 we can write

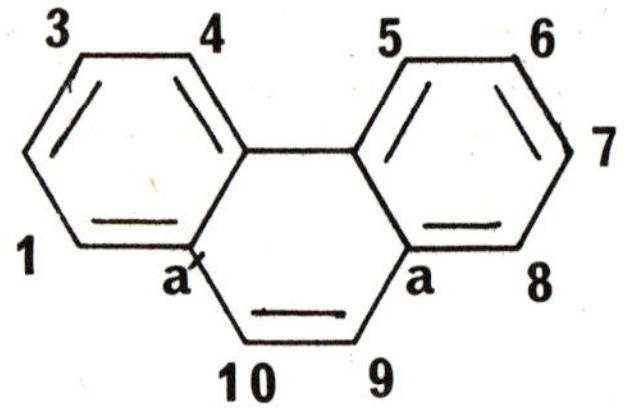

$$\Delta E_o = -2(p_{a9} + p_{9\ 10} + p_{a'\ 10})$$

By symmetry p_{a9} equals $p_{a'\ 10}$.

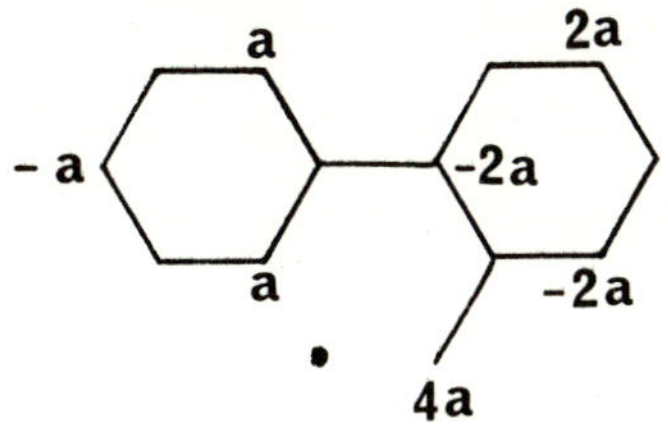

$$31a^2 = 1;\ a = 1/\sqrt{31} = 0.179$$

$$p_{a9} = p_{a'\,10} = 1 \times 0.179 = 0.179$$

$$p_{9\,10} = 1 \times 0.716 = 0.716$$

and

$$\Delta E_o = -2(2 \times 0.179 + 0.716)\,\beta = -2.15\,\beta$$

The ortholocalization energy is usually expressed as a positive number (i.e., in units of $-\beta$).

When the bond is not as symmetrically located it is clear that one might calculate more than one bond order for the value between the carbons where localization is to occur. Let us consider the 1-2 bond in phenanthrene.

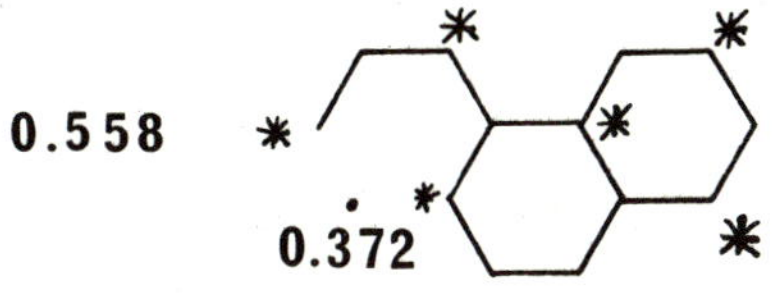

$p_{a\,1} = 0.372 \qquad p_{12} = 0.558$

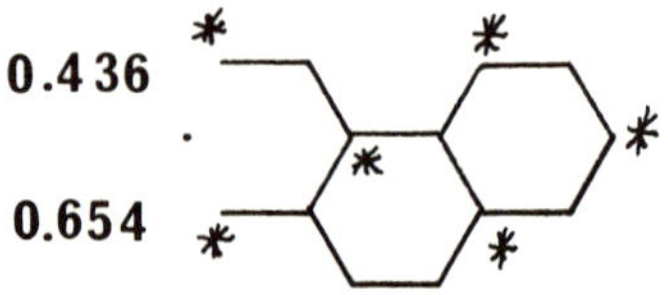

$p_{12} = 0.654 \qquad p_{23} = 0.436$

To calculate ΔE_o the average value for p_{12} is used.

$$\Delta E_o = -2(0.372 + \frac{1}{2}(0.558 + 0.654) + 0.436)\ \beta$$

$$= -2.828\ \beta$$

The bond with the lowest ortholocalization energy is the 9-10 bond (2.15), and it is here that the reaction occurs. A study of reactive sites in a substantial number of aromatic hydrocarbons has shown an excellent correlation.

Carcinogenic Activity

Some years ago the French research team of B. Pullman and wife set about studying the electronic properties of those polynuclear hydrocarbons which were known to be capable of inducing cancer. MO calculations on these substances showed regions of olefin-like bond character designated as K regions.

Subsequently, it was shown that the character of various transannular regions (L regions) was also required to characterize carcinogenic potentials. The two regions are shown here for benzanthracene (which incidently is not carcinogenic).

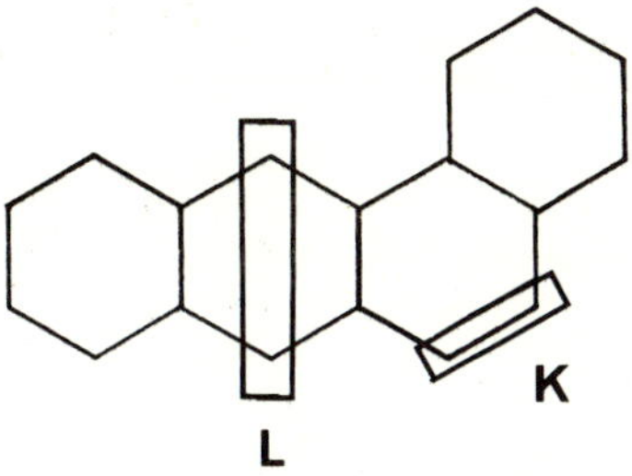

The paralocalization and ortholocalizations approximated from PMO theory may be used for the characterization. The requirement for carcinogenic character appears to be ortholocalization energies below about 2β providing there is no region of paralocalization energy of less than about 2.9β.

Two examples of particularly potent carcinogens are

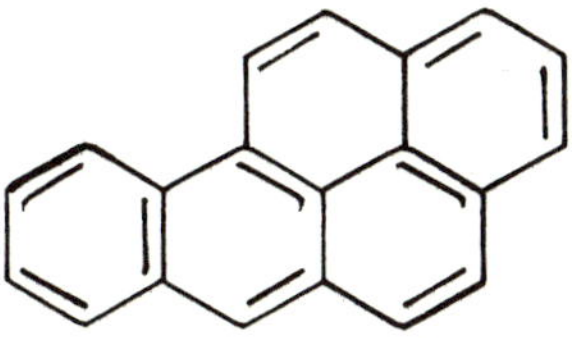

3,4-Benzopyrene

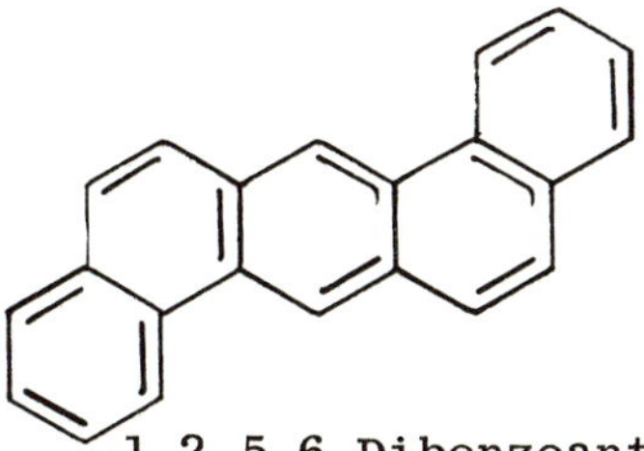

1,2,5,6-Dibenzoanthracene

Exercises

1. Calculate N_t for the various carbons in phenanthrene and show that these are roughly in accord with the relative order of nitration, i.e., with benzene as 1 the values for the various positions in phenanthrene are (1) 360, (2) 92, (3) 300, (4) 79, and (9) 490.

2. Calculate the expected position(s) of electrophilic attack in quinoline and in isoquinoline.

3. Compare the ortholocalization energies of 3,4-benzopyrene with those of benzene and naphthalene, and with benzanthracene.

4. Show that the resonance theory and PMO theory lead to different conclusions regarding the position of electrophilic substitution in biphenylene. The PMO method gives the correct answer in this case.

5. Account for the fact that pyridine undergoes nucleophilic substitution by amide ion to form either 2-amino or 4-aminopyridine in preference to 3-aminopyridine.

6. Predict the position of attack by dienophils on phenanthrene.

Chapter 8

PERICYCLIC REACTIONS

As evidence regarding organic reaction mechanisms was acquired and digested during the earlier decades of this century, several types of reactions became conspicuous by virtue of the absence of any indications of reactive intermediates. For the most part, these cases had in common their rather exact stereospecificity and the absence of any catalysts other than the input of energy in the form of heat and light.

To some it appeared that these reactions might involve cyclic, neutral transition states, which were depicted as being formed by electrons moving along the paths of curved arrows, i.e.,

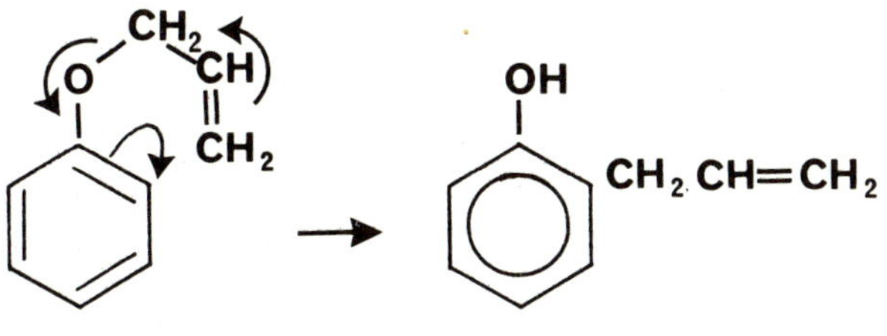

(8-1)

Name reactions such as the Diels-Alder, the Cope rearrangement, and the Claisen rearrangement fell in this category.

Recently, it has become customary to use more exacting definitions descriptive of the reaction type. The three current classifications are exemplified by the following:

Electrocyclic reactions

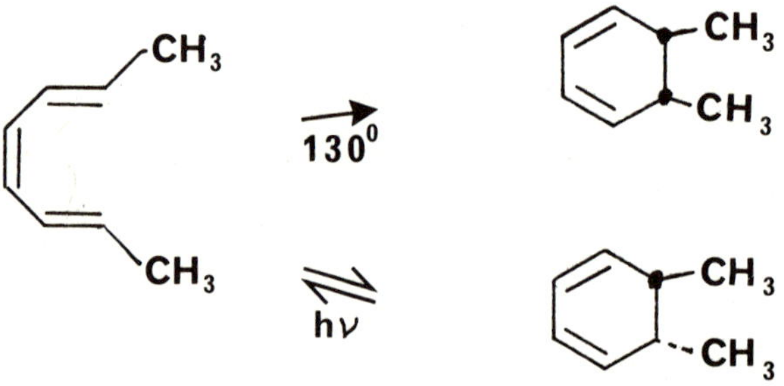

(8-2)

Cycloaddition reactions

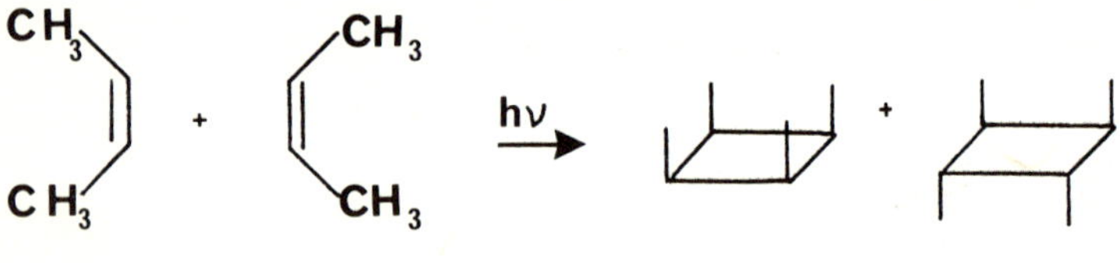

(8-3)

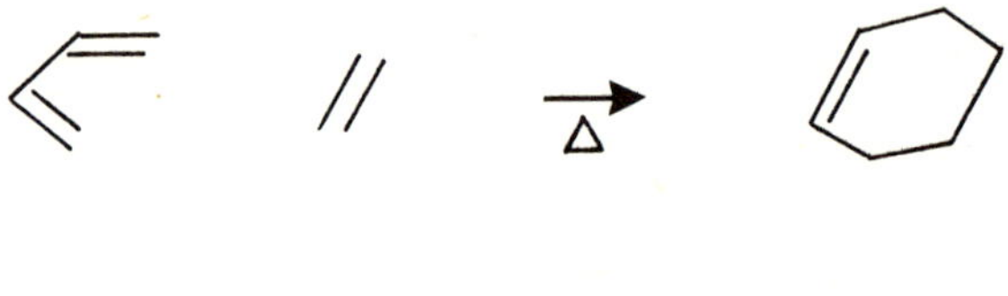

(8-4)

Sigmatropic reactions

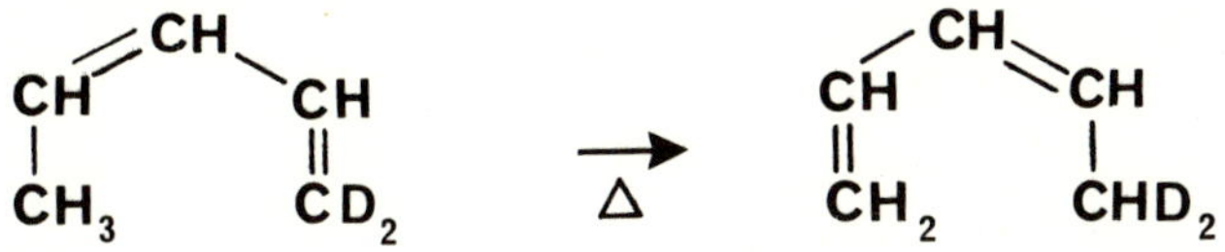

(8-5)

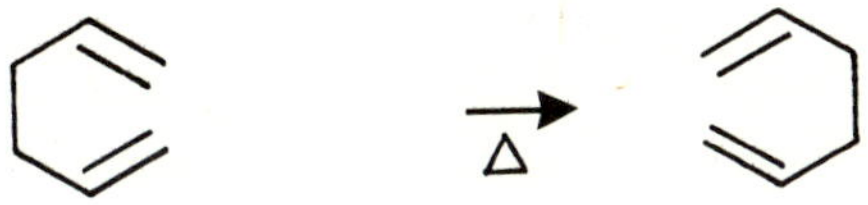

(8-6)

It will be obvious to organic chemists that the Diels-Alder reaction is a cycloaddition process, while the Cope and Claisen rearrangements are sigmatropic reactions.

Failure to produce concrete mechanistic evidence over the years proved frustrating to the point where it was once "a la mode" to refer to reactions of this type as "no mechanism" reactions. In 1965, however, all of this changed with the announcement of a series of rules by Professors Woodward and Hoffmann designed to predict the ease and steric course of all such concerted reactions. Indeed, the whole category was lumped into a more general type called "pericyclic reactions" proceeding through cyclic transition states formed by the continuous making of new bonds and breaking of the old.

The powerful predictive utility of the Woodward-Hoffmann rules has been thoroughly verified in the ensuing years, and even now the chemical literature is filled with a continuing spate of new and imaginative research suggested by the rules.

Woodward and Hoffmann originally grounded their rules in the so-called "frontier orbital" method and an extended form of HMO calculations. Subsequently, Longuet-Higgins and Abrahamson produced an argument supporting the rules based on the symmetry properties of the ground and transition state wave functions. Woodward and Hoffmann also utilized an orbital symmetry justification which will be described in more detail later.

Evans' Rule, PMO Theory, and Pericyclic Reactions

During the period marking the onset of World War II M. G. Evans in England published several papers dealing with the mechanism of the Diels-Alder reaction. In these it was pointed out that the reaction was probably concerted and that a cyclic transition would possess the same stabilizing influence found in aromatic systems. Due to the war Evans' work was largely ignored until it was returned to light by Professor Dewar.

The concept of an aromatic transition state is particularly easy to deal with using PMO theory, for as we have seen in Chapter 6, the energy differences between cyclic and linear systems are readily ascertained. Dewar has generalized Evans' results to all pericyclic reactions, but before stating the generalization let us consider the case of the reaction of butadiene with ethylene (8-4).

The formation of cyclohexene as shown in (8-4) requires that butadiene and ethylene collide to form some sort of a transition state which then collapses, forming the products. Two possibilities exist, which are diagrammed below.

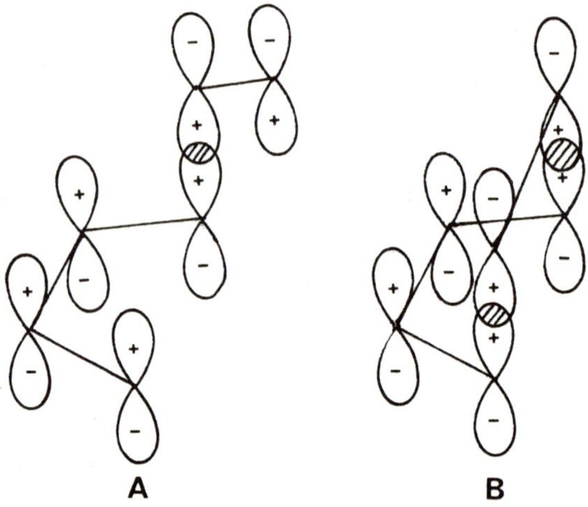

(8-7)

Several points may be made regarding the above diagrams. For one thing, we will continue to approximate π-MOs as a linear combination of p-AOs. The lobes of each AO are opposite in sign. In choosing the basis sets as above, it is customary to choose the AOs so that they overlap with the same sign or phase, positive with positive, etc. This choice is arbitrary, but it can be shown that the resultant MOs are independent of the signs of the basis set. It may be well to reiterate, however, that in (8-7) above and in subsequent diagrams the signs shown are associated with the basis set AOs.

Examination of A and B above allows a comparison with hexatriene and benzene respectively. For hexatriene one has six AOs and six π-electrons arranged in an open chain fashion, while in benzene the six AOs and π-electrons comprise a cyclic system. Two systems are said to be <u>isoconjugate</u> if they have the same number of π-electrons occupying π-MOs of the same size and shape and formed by the interaction of the same number of AOs. Thus, A is isoconjugate with linear hexatriene, and B is isoconjugate with benzene. As shown previously (p. 83), hexatriene is less stable than benzene, and rationally the preferred transition state for the Diels-Alder will be the cyclic form B. The Diels-Alder reaction is an example of a (4 + 2) cycloaddition reaction.

The thermal dimerization of ethylene to form cyclobutane would be a (2 + 2) addition which might similarly be expected to go through either a linear or cyclic transition state.

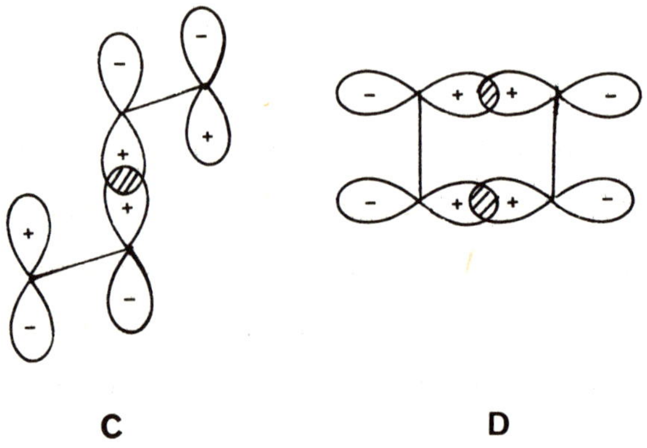

(8-8)

The cyclic transition state D is isoconjugate with cyclobutadiene and, therefore, antiaromatic. Such dimerizations do not occur readily, and where they do a two step diradical process is indicated.

With the above examples behind us it is now proper to give a more complete statement of Evans' Rule, namely,

> Thermal pericyclic reactions take place preferentially through aromatic transition states.

It should be added that the operating assumption in thermal reactions is that they occur with the reactant(s) in the ground state electronic configuration. More details must be supplied before Evans' Rule can be extended to photochemical reactions.

Anti-Hückel Systems

Before proceeding to a more detailed consideration of pericyclic reactions it is necessary to explore farther the changes produced in alterating the signs given the basis set AOs.

As mentioned above, conventionally the AOs in ethylene are chosen to be in phase and the resonance integral is assigned the value β.

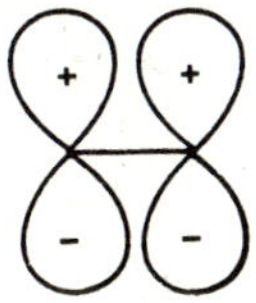

β

If one of these AOs is now twisted about the bond axis, the resonance integral will vary as the cosine of the twist angle. At 180° the value becomes $-\beta$.

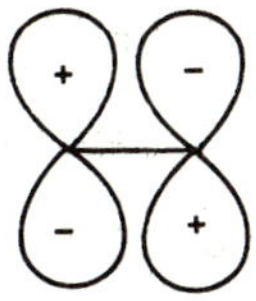

$-\beta$

Consideration of this effect could produce a profound alteration in π-energies for polyenes calculated by the PMO method for NBMO coefficients change sign with β. But does it?

Consider for a moment the effect of phase dislocations introduced by twisting either an internal or terminal AO in pentadienyl.

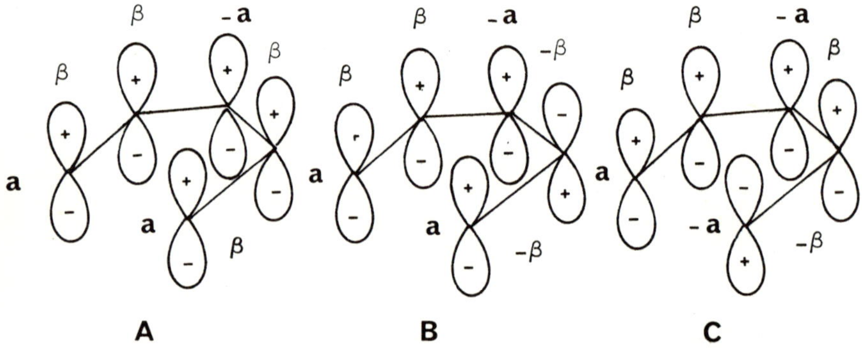

(8-9)

The first case (A) is the usual one. With the internal dislocation (B) two resonance integrals are reversed but the NBMO coefficients remain the same. Union of methyl with B will give the same results as with (A). Furthermore, in the case of (8-9C) union with methyl to form benzene would give $\delta E = [2\ (a)\ (\beta) + (-a)\ (-\beta)] = 4a\ \beta$, which again is the value obtained from (A) and (B). Thus, as long as the number of changes in β

around the ring is zero or an even number, the usual criteria for aromaticity apply.

While no stable ring systems containing an odd number of changes in β can exist, it is interesting to consider how such a system might arise. Heilbronner has pointed out that an annulene with a half twist in the ring would create a Mobius strip molecule with one phase dislocation.

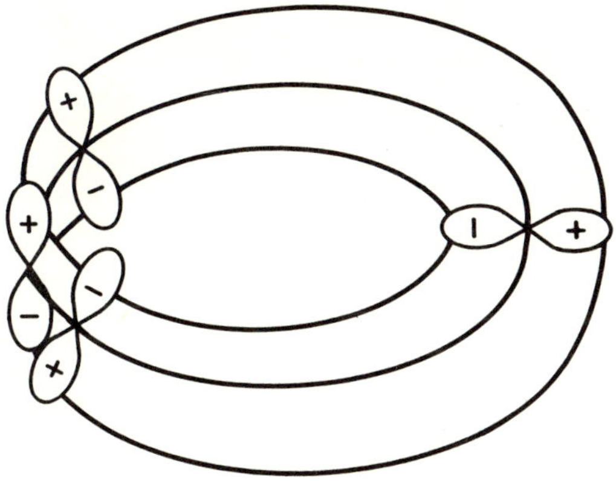

(8-10)

Considerations of the change in π-energy on forming such a system reveal that now the (4n + 2) systems are antiaromatic, while those with 4n π-electrons are aromatic. Systems with an odd number of phase dislocations have been dubbed <u>anti-Hückel</u> systems. Their importance appears in transition states for certain pericyclic reactions.

A summary of what to expect from Hückel and anti-Hückel systems is given here, where (+) denotes aromatic and (-) antiaromatic character.

	Rings					
	Even-numbered		Odd-numbered			
	4n	4n+2	4n+1		4n+3	
			Cations	Anions	Cations	Anions
Hückel	-	+	-	+	+	-
anti-Hückel	+	-	+	-	-	+

Electrocyclic Reactions

Electrocyclic reactions, as illustrated in (8-2), involve the interconversion of a conjugated polyene and the cyclic isomer formed by union between the terminal carbons. Since the reactions are usually reversible, we may apply the principle of microscopic reversibility, and for convenience of discussion consider a ring opening reaction.

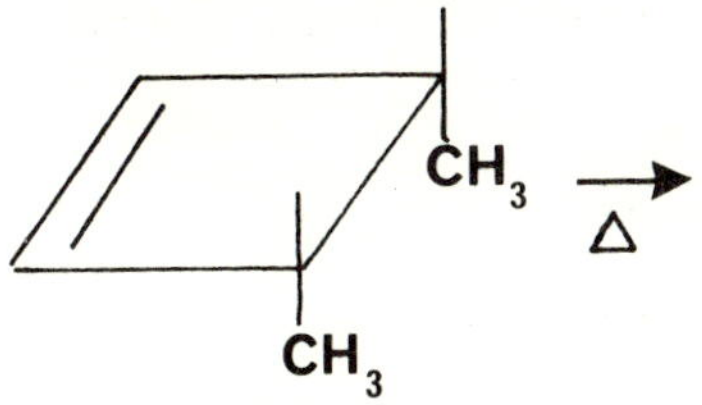

$CH_3CH{=}CHCH{=}CHCH_3$ (8-11)

Given that only one product is formed in a stereospecific fashion, it is evident that one of three geometrical isomers must result, i.e., cis, cis (A), trans, trans (B), and cis, trans (C).

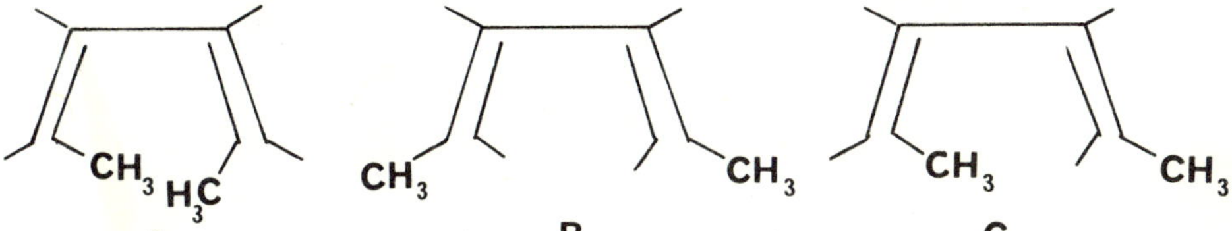

A B C

Clearly, the ring opening must be accompanied by rotation of the methyl groups into their appropriate planar configurations. Woodward and Hoffmann have termed the two possible modes <u>disrotatory</u> and <u>conrotatory</u> depending on the sense of the rotation about the developing double bonds.

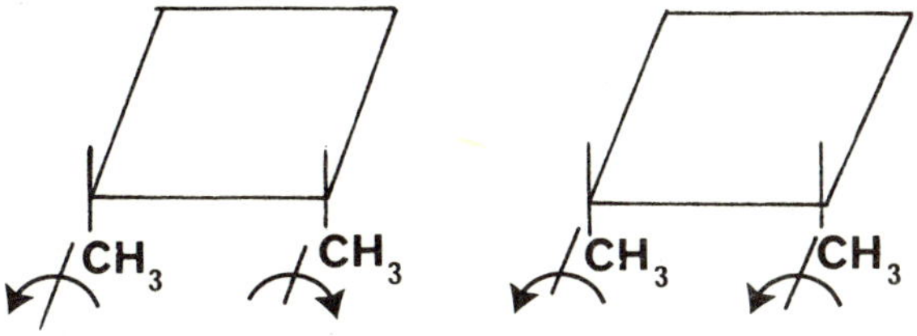

(8-12)

Disrotatory Conrotatory

Examination shows that (A) and (B) can arrive only *via* disrotatory processes, while (C) results from a conrotatory process.

The transition states for the ring opening will require the two carbons bearing the methyls to develop sp^2 character as the sigma bond ruptures. The transition state will be isoconjugate with cyclobutadiene with four *p* orbitals and four π-electrons.

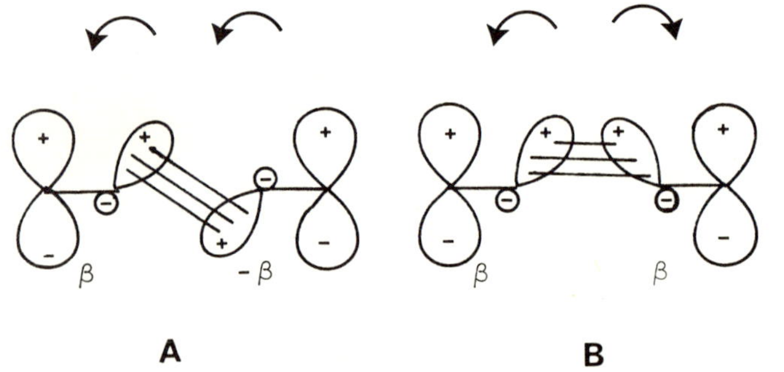

(8-13)

Conrotatory opening   Disrotatory opening

In the case of disrotatory opening a normal Hückel system will be developing so that the transition state will be antiaromatic. However, for the conrotatory process an anti-Hückel system with one phase dislocation is formed, and this will be aromatic. Obviously the conrotatory process is favored, and one does isolate only the trans, cis product (C) from the reaction.

Consideration of the two possible transition states for the thermal ring closure in (8-2) shows the disrotatory process to lead to a transition state isoconjugate with benzene.

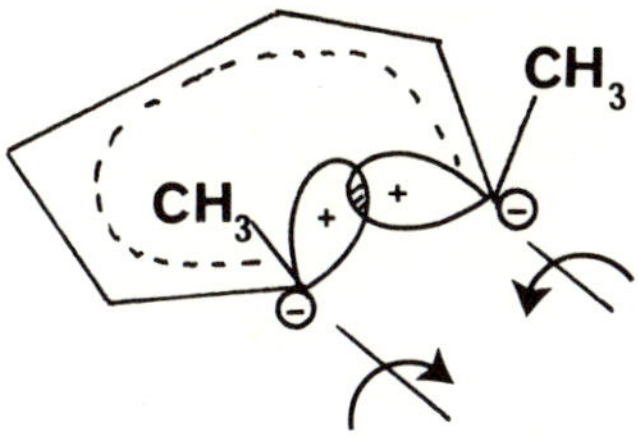

(8-14)

Thus, thermal cyclizations of 1,3,5-trienes will preferentially occur _via_ the aromatic (4n + 2) transition state resulting from the disrotatory process.

Photochemical Reactions

As demonstrated in (8-2), pericyclic reactions may take place through the action of light. Often the products are different from those produced thermally. Usually this difference appears in the stereochemistry of the product.

It is now well known that photochemical reactions occur through excited reactants; singlets and triplets formed by promoting an electron from the highest occupied orbital of the ground state to the lowest unoccupied orbital.

The PMO method is not capable of distinguishing singlet and triplet states. However, this fact appears to bear no consequences for qualitative arguments on photochemical pericyclic reactions.

While we will not do so here, it is possible to make use of arguments similar to those presented previously regarding the stabilities of transition states to show that photochemical reactions proceed in just the opposite sense of thermal reactions, i.e.,

> Photochemical pericyclic reactions take place preferentially through excited forms of antiaromatic transition states.

This means that the conclusions regarding the permissiveness and stereochemistry of photochemical reactions will be just the reverse of the corresponding thermal reactions.

Thus, the (2 + 2) cycloaddition reactions of olefins to form cyclobutanes, while thermally disallowed, are photochemically allowed. Indeed, the photochemical pathway is often the synthetic method of choice in the preparation of substituted cyclobutanes [see (8-3)].

For the photochemical electrocyclic reaction (8-2) the (4n + 2) atoms of the transition state must show a phase dislocation since in order to be antiaromatic it must also be anti-Hückel (see Table 8-1). Hence, the transition state will look like:

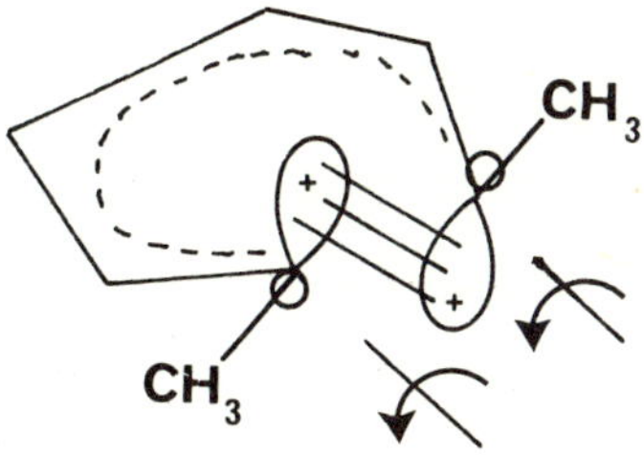

(8-15)

Sigmatropic Rearrangements

These reactions are defined by the migration of a sigma bond flanked by one or more π-bonds to a new position in the molecule.

$$\overset{X}{C}\,(C{=}C)_n \;\text{———————}\; (C{=}C)_n\,\overset{X}{C} \qquad (8\text{-}16)$$

In addition to the examples (8-5) and (8-6) given previously, two other examples will be given here.

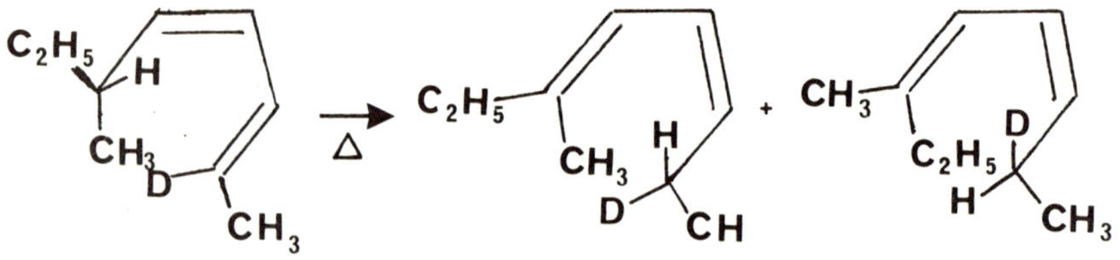

(8-17)

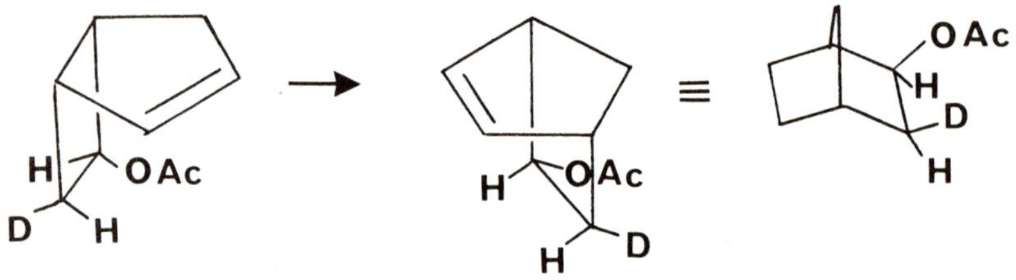

(8-18)

It will be more convenient if two matters of nomenclature are cleared up before proceeding. First, the reaction is termed an [i, j] sigmatropic rearrangement when the bond migrates from position [1, 1] to position [i, j]. Reaction (8-5) is a [1, 5] rearrangement, while (8-6), (8-17), and (8-18) are [3, 3], [1, 5] and [1, 3], rearrangements respectively.

Secondly, if the migrating group remains on the same side of the π-system the process is said to be suprafacial. Migrations which cross the nodal plane of the π-system are called <u>antarafacial</u>.

Consider the transition states for the [1, 3] rearrangement of hydrogen. The structure (A) is isoconjugate with cyclobutadiene and, as a Hückel system, is antiaromatic and unstable. While the antarafacial, anti-Hückel transition state (B) is energetically favorable, the span for H orbital overlap is prohibitively large.

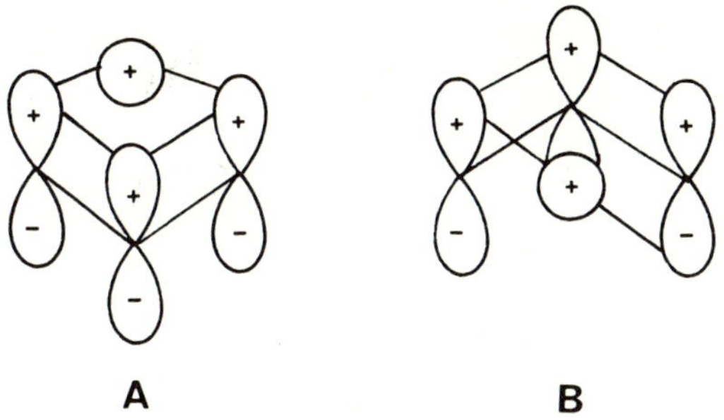

(8-19)

Thus, thermal [1, 3] rearrangements of hydrogen are not observed.

However, in (8-18) the rearrangement may proceed through the anti-Hückel system (8-20), which is not only aromatic but sterically possible.

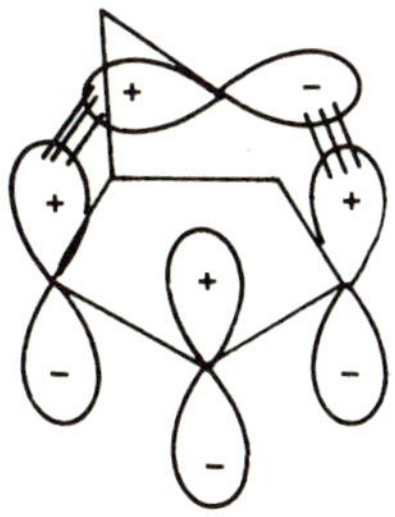

(8-20)

The [1, 5] rearrangements of (8-5) and (8-17) proceed through suprafacial, Hückel (4n + 2) transition states.

Summary

The PMO approach when combined with Evans' rules allows facile predictions regarding pericyclic reactions. One need only examine possible transition states and decide which resemble Hückel or anti-Hückel systems. It is then a simple matter of applying Table 8-1 in deciding which will be aromatic or antiaromatic.

The Conservation of Orbital Symmetry

The basis behind the approach of Woodward and Hoffmann is the concept that the MOs of the product must be generated smoothly from MOs in the reactants having the same symmetry properties. Thus, for a thermal pericyclic reaction the electrons of the reactant(s) will be in the ground state MOs. If these MOs are converted into MOs of the product without a change in symmetry the reaction is allowed, i.e., energetically favorable. For photochemical reactions the reactant is in the first excited state. Again the differentiation between singlet and triplet states is ignored.

Several approaches to the Woodward-Hoffmann rule have been made. The most popular and utilitarian appears to be the use of correlation diagrams, and as a comparison to the PMO approach we will briefly examine how these are drawn. Briefly, the task is to write down the approximately known energy levels of the reactants on one side and the products on the other. Assuming a geometry of approach, one classifies both sets according to the symmetry maintained during the approach. Levels of like symmetry are then connected following the rule that only levels of unlike symmetry are allowed to cross.

Consider again the union of two ethylenes to form cyclobutane. In all, four MOs are involved (a bonding and an antibonding for each ethylene) for the reagents, and similarly there will be four new sigma orbitals in the product. The Woodward-Hoffmann rule requires that for the reaction to occur in a concerted fashion the orbitals of the reactants must pass into those of the products in a continuous manner, i.e., the symmetry of the orbitals must be conserved.

The approach of two ethylenes to form the cyclic transition state is characterized by two symmetry planes as diagramed below:

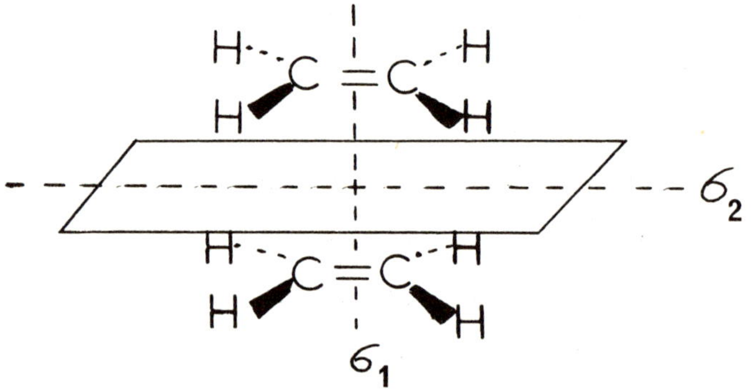

(8-21)

The interaction of the two sets of ethylene MOs can be expressed in terms of these symmetry planes and the phases of component atomic orbitals.

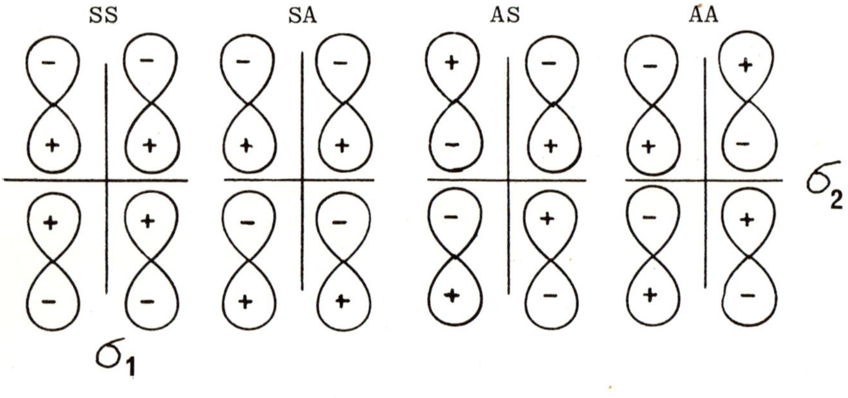

(8-22)

The designations indicate whether the orbitals are symmetric (S) or antisymmetric (A) about the symmetry planes σ_1 and σ_2, respectively. If the wave functions are taken as ψ_1 and ψ_2 for the ground states and ψ_1^* and ψ_2^* for the antibonding states, then the wave functions for the reactants in the above diagrams can be written as

$$\Psi_1 = 1/\sqrt{2}\ (\psi_1 + \psi_2) \qquad \text{SS}$$

$$\Psi_2 = 1/\sqrt{2}\ (\psi_1 - \psi_2) \qquad \text{SA}$$

$$\Psi_3 = 1/\sqrt{2}\ (\psi_1^* + \psi_2^*) \qquad \text{AS}$$

$$\Psi_4 = 1/\sqrt{2}\ (\psi_1^* - \psi_2^*) \qquad \text{AA}$$

Since the π-systems are independent in the reactants, E_1 and E_2 are equal as are E_3 and E_4.

In progressing to the new sigma bonds of the product the symmetry must be preserved under the rule.

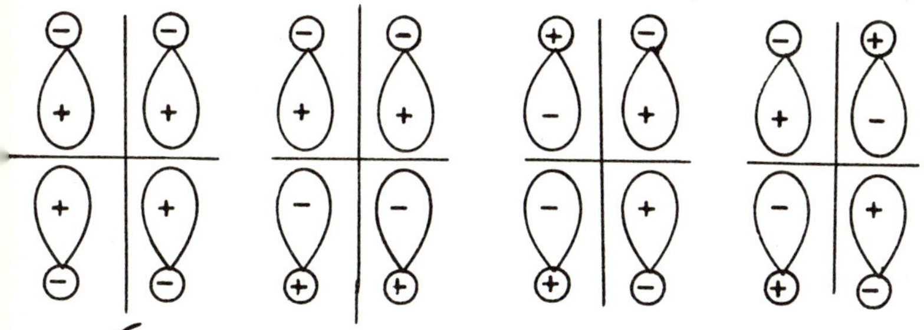

σ_2 (8-23)

σ_1

New wave functions for the two sigma bonds may be defined as

$$\Phi_1 = 1/\sqrt{2}\ (\phi_1 + \phi_2) \qquad \text{SS}$$

$$\Phi_2 = 1/\sqrt{2}\ (\phi_1 - \phi_2) \qquad \text{SA}$$

$$\Phi_3 = 1/\sqrt{2}\ (\phi_1^* + \phi_2^*) \qquad \text{AS}$$

$$\Phi_4 = 1/\sqrt{2}\ (\phi_1^* - \phi_2^*) \qquad \text{AA}$$

where Φ represents the newly formed sigma bond wave functions formed from carbon sp^3 atomic orbitals. Since the two sigma bonds are independent in the product, again E_1 will equal E_2 and E_3 will equal E_4. We can now plot the relative energies of the various orbitals as follows

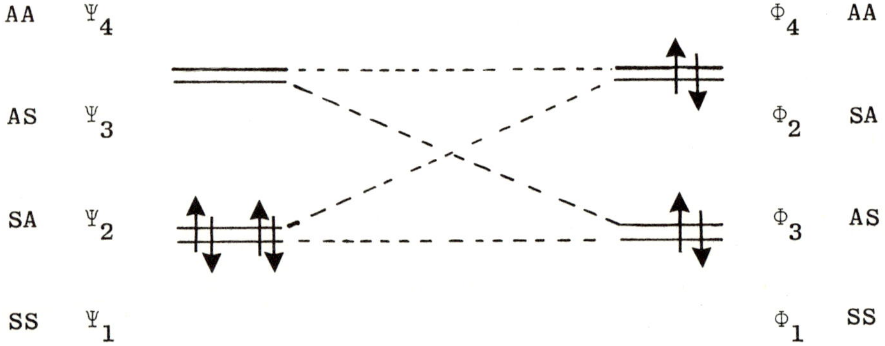

Note that it is Φ_3, an AS function, which becomes a bonding MO in the product, while the MO associated with Φ_2 is antibonding. It follows under the Woodward-Hoffmann rule that for a concerted (2 + 2) cycloaddition where orbital symmetry is

preserved a high energy transition state leading to an electron pair in an antibonding orbital of the product is formed. Such a process is considered energetically forbidden. If, however, one of the reagent electrons is promoted to Ψ_3 then the formation of a diradical product becomes thermochemically allowed. Such a promotion can be brought about by the absorption of a photon.

An entirely analogous process may be carried out for a (4 + 2) cycloaddition reaction such as the butadiene plus ethylene reaction. The same symmetry elements exist for this system as for that above. The correlation diagram is as follows

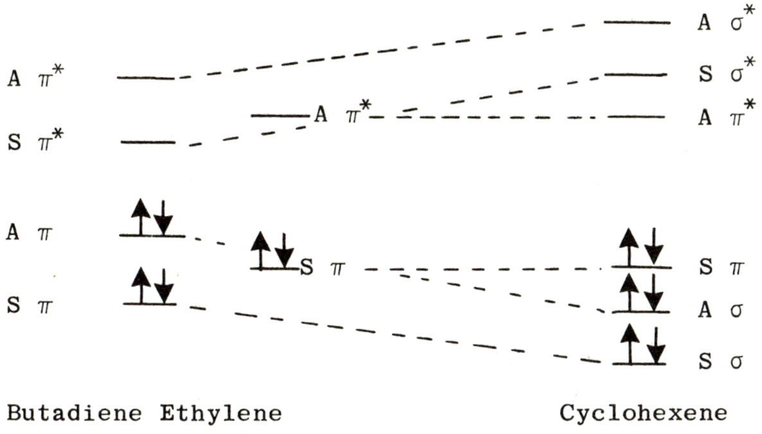

Here symmetric and antisymmetric π-orbitals are

designated Sπ and Aπ with the stars reserved for the antibonding MOs. Similar notation is used for the sigma bonds in the product. The MOs for ethylene and butadiene were given previously on pages 18 and 21. Since the preservation of orbital symmetry leads only to a net lowering of energy this process is thermochemically allowed.

Electrocyclic Reactions

In order to apply orbital symmetry considerations to electrocyclic reactions it is necessary to consider the consequences of both conrotatory and disrotatory processes. Again let us consider the interconversions of cyclobutenes and butadienes. It will be necessary to consider the rupture of the cyclobutene sigma bond and the π-system as they are converted into butadiene orbitals. Let us consider these aspects independently with reference to the symmetry axis of the molecule and for the conrotatory process.

S σ

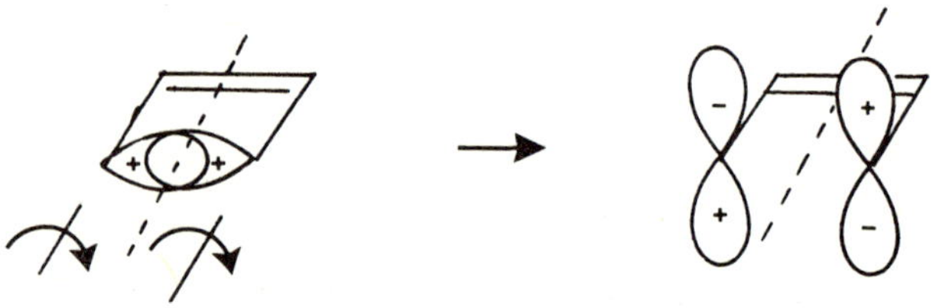

S ψ_2 or ψ_4

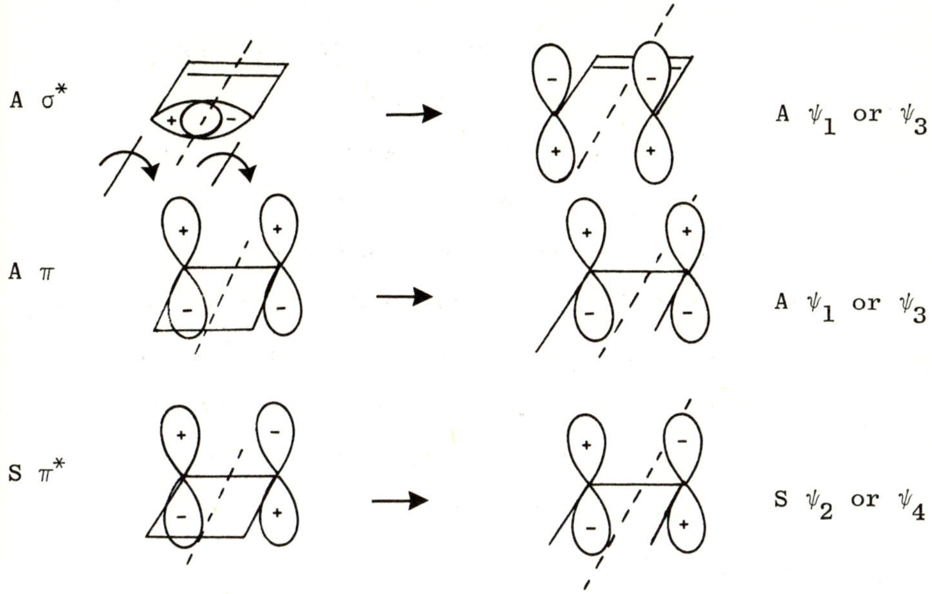

The symmetry designation refers to the symmetry axis shown and those butadiene orbitals which correspond to the symmetry of the AOs shown. An energy correlation diagram may now be drawn.

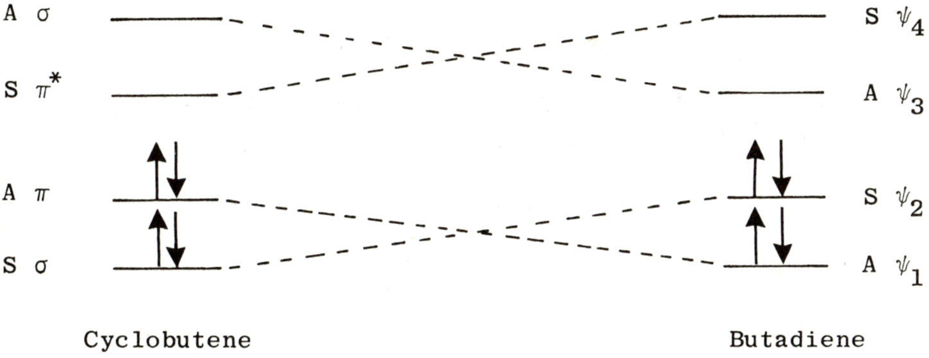

The conrotatory ring opening is seen to be thermochemically allowed. Essentially the same process may be followed to show that the disrotatory process is forbidden, the symmetry element in this case being a plane.

The point regarding the construction of correlation diagrams and the application of the rule of conservation of orbital symmetry is made in the above examples, and more detailed texts on the subject should be consulted for the extension to sigmatropic rearrangements.

Conclusion

Which approach is better in treating pericyclic reactions? In most cases the PMO and Woodward-Hoffmann approaches are in complete agreement, and the practitioner is left to choose that scheme with which he is most comfortable. The author has felt that the PMO approach was generally easier to learn and apply, but others may disagree. Dewar has produced an argument that the PMO theory is on a more sound theoretical basis than the foundations offered for the Woodward-Hoffmann approach.

Exercises

1. What stereochemistry would you expect from the thermal ring opening reactions of cis and trans 5,6-dimethyl-1,3-cyclohexadiene?

2. Demonstrate for each of the products in (8-17) whether or not the hydrogen migrates in a suprafacial or antarafacial manner.

3. Discuss the mode of migration in the following cases and show them to be consistent with the PMO approach.

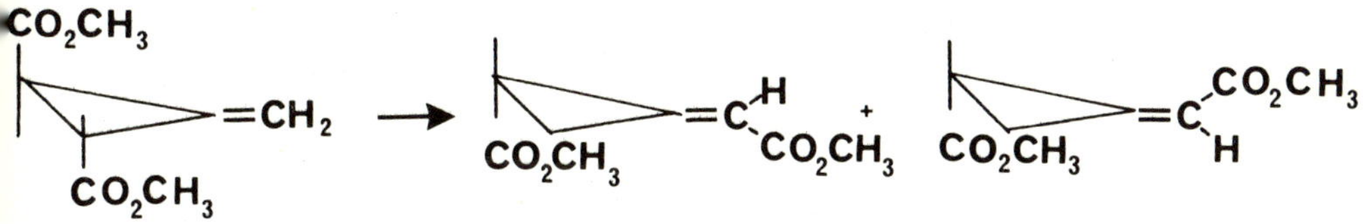

4. Upon heating, the deuterated compound on the left suffers scrambling. Since one can not separate the isomers, the effect is as though there were multiple labeling as indicated in the equation. Account for this observation.

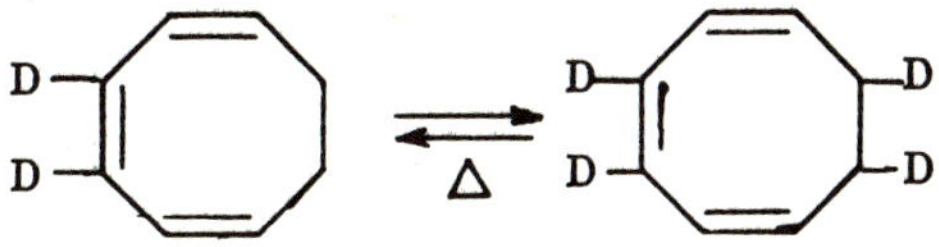

5. The Cope rearrangment of biallyl below may be pictured as proceeding via either a chair or boat transition state. Which would you favor and why? [Note: * denotes C-14.]

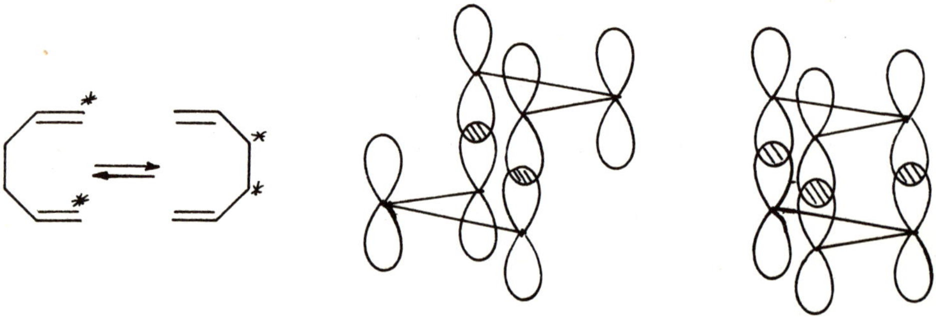

BIBLIOGRAPHY

Chapters 1 - 4

For more detailed account of the Hückel method see the following.

Liberles, A., Introduction to Molecular-Orbital Theory, Holt Rinehart Winston, New York, 1961.

Roberts, J. D., Notes on Molecular Orbital Calculations, W. A. Benjamin, New York, 1961.

Streitwieser, A., Molecular Orbital Theory for Organic Chemists, Wiley, New York, 1961.

Chapter 5

A readable treatment of perturbation is given by:

Anderson, J. M., Introduction to Quantum Chemistry, W. A. Benjamin, Inc., New York, 1969.

For the fundamental papers on PMO theory, see:

Coulson, C. A. and H. C. Longuet-Higgins: Proc. Roy. Soc. (London), A191, 39 (1947); A192, 16 (1947); A193, 447, 456 (1948); A195, 188 (1948).

Dewar, M. J. S., J. Amer. Chem. Soc., 74, 3341, 3345, 3350, 3353, 3357 (1952).

Longuet-Higgins, H. C., J. Chem. Phys., 18, 265, 275, 283 (1950).

Other textbook accounts are given in:

Dewar, M. J. S., The Molecular Orbital Theory of Organic Chemistry, McGraw-Hill, New York, 1969.

Flurry, R. L., Molecular Orbital Theories of Bonding in Organic Molecules, Dekker, New York, 1968.

Riggs, N. V., Quantum Chemistry, McMillan, Toronto, 1969.

Chapter 6

See Dewar and Riggs under Chapter 5.

Also see:

Dewar, M. J. S., Tetrahedron, Suppl. 8, Part 1, 75 (1966).

Dewar, M. J. S. and H. C. Longuet-Higgins, Proc. Roy. Soc., A214, 482 (1952).

Hess, B. A. and L. J. Schaad, J. Amer. Chem. Soc., 93, 305 (1971).

Herdon, W. C., Enumeration of Resonance Structures, Tetrahedron, 29, 3 (1973).

Platt, J. R., Encyclopedia of Physics, (S. Flügge, Ed.), Spinger-Verlag, Berlin, 1961, p. 171.

Wilcox, C. F., Tetrahedron Letters, 795 (1968).

Chapter 7

See references for Chapter 5.

Chapter 8

Dewar, M. J. S., Angew. Chemie. (English Ed.), 10, 761 (1971).

INDEX

A

Acridine, 65
Allyl, 28, 84
Allyl phenyl ether, 127
o-Allylphenol, 127
Alternant systems, 39
Amine, base strengths, 43, 63
1-Aminoacridine, 67
9-Aminophenanthridine, 67
4-Aminopyridine, 65
4-Aminoquinoline, 65
Aniline, 43
Antarafacial migrations, 144
Antiaromaticity, 75, 84
Antibonding MOs, 15
Anti-Hückel systems, 135, 137, 145
Aromaticity, 75, 80
Atomic orbitals, 6
Azacyclopropene, 53, 56
Azulene, 89, 100, 103

B

Benzene, 33, 80, 83, 111
3,4-Benzopyrene, 125
Benzyl, 41, 72, 93, 101
  charge densities, 43
  NBMO coefficients, 41
Biphenyl, 87, 103
Biphenylene, 87, 98
Bohr atom, 1
Bond, essential, 81, 87
  integral, 10, 18
  lengths, 36
  localized, 77, 80
  order, 35
Bonding MOs, 15
2-Butene dimerization, 128
Butadiene, 21-28, 36, 68, 78, 79, 151

C

Carcinogenic activity, 124
Charge density, 38
4-Chloronaphthalene, 119
Claisen rearrangement, 129
Coefficients, wave function, 25
Conrotatory motion, 139
Cope rearrangement, 129
Coulomb integral, 9, 18
  perturbation effects on, 60
Coulomb potential, 5
Coulson, C. A., 35, 55
Cross Conjugation, 34
Cycloaddition reactions, 128, 142, 151
Cyclobutadiene, 34, 84, 97, 145
Cyclopropenyl, 30
Cyclooctatetraene, 98, 103

D

De Broglie, L., 2
Degenerate orbitals, 30
Delocalization energy, 24, 108
Determinants, secular, 12
  reduction, 21
Dewar, M., 55, 73, 77,

109, 122, 131
1,2,5,6-Dibenzoanthracene, 125
Diels-Alder reaction, 119, 129, 133
Dipole moments, 44-46
Disrotatory motion, 139

E

Eigenfunctions, 2
Eigenvalues, 2
Electrocyclic reactions, 128, 138, 143, 152
Electron density, 38
Electrophilic substitution, 107
Ethylene, 18-20, 133, 147, 151
Evans' rule, 130, 134, 142
Even AHC, 39

F

Formaldehyde, 53
Free radical substitution, 115, 117
Fulvene, 46, 81, 102

H

Hamiltonian, 2
Hammond's postulate, 108
Heilbronner, E., 137
Hess, B., 79, 82, 102
Heteroatom systems, 49-53, 111
2,4-Hexadiene, 139
1,5-Hexadiene cyclization, 129
Hexatriene, 34, 74, 83
HMO theory, 17-34
Höffmann, R., 129, 130
Hückel MO theory, 17-34
Hückel rule, 85
Hückel systems, 145
Hydrocarbons, alternant, 39
aromatic, 75-80
nonalternant, 39
Hydrogen atom, 5
Hydrogen molecule, 6
energy levels, 13
wave functions, 14

I

Isoconjugate systems, 133
Isoquinoline, 64

K

Kekule, A., 75
K regions, 125

L

LCAO, 8
Localization energy, 108
Localized bond, 77, 80
Longuet-Higgans, H. C., 55, 67
L regions, 125

M

Method of Variations, 8
Methylenecyclopropene, 45
Methylene imine, 52
α-Methylenenaphthalene, 42
Mobius strip molecules, 1 137
MO coefficients, 25
Molecular orbitals, 7

N

Naphthalene, 89, 100, 110, 111, 121
NBMO, 28, 41
Nonalternant hydrocarbons, 39, 88
Nonbonding MOs, 28, 41
Normalization, 5
Normalization factor, 6
Nucleophilic substitution, 115, 118

O

2,4,6-Octatriene cyclization, 128
Odd AHC, 39
Operators, 2
Orbital symmetry, 146
Orthogal functions, 11
Ortholocalization energy, 121
Osmate esters, 121
Overlap integral, 10

P

Paralocalization energy, 119
Parity, 81
Pauling, L., 35, 76
Pentadienyl, 83, 103, 136
Pericyclic reactions, 127-156
Perturbation theory, 55
  first order, 56, 71
  second order, 70
Petit, R., 84
Phenanthrene, 102, 121
Phenol, 54
1-Phenyl-1,3,5-hexatriene, 88
Photochemical reactions, 141
PMO method, 55-73, 80
Polynomial roots, 22
Pullman, B., 124
Pyridine, 52, 64, 113, 115

Q

Quinoline, 64

R

Reaction rates, 105
Reactivity numbers, 109
Relative reactivities, 106
Resonance, 75, 91-102
Resonance integrals, 10, 18
Robinson, R., 76

S

Schaad, L., 79, 82, 102
Schrödinger, E., 2
Secular determinant, 12
Secular equation, 12
Sigmatropic reactions, 129, 143
Starring, 39
Styrene, 72, 103
Supernumerary MOs, 98
Suprafacial migrations, 144

U

Union of π-systems, 68, 82, 87

V

Variations, method of, 8

W

Wave equation, 2
Wave function, 2
Wheland, G., 76, 108
Woodward-Hoffmann rule, 130, 147
Woodward, R., 129, 130

Other books of interest to you...

Because of your interest in our books, we have included the following catalog of books for your convenience.

Any of these books are available on an approval basis. This section has been reprinted in full from our ***inorganic and organic chemistry*** catalog.

If you wish to receive a complete catalog of MDI books, journals and encyclopedias, please write to us and we will be happy to send you one.

MARCEL DEKKER, INC.
95 Madison Avenue, New York, N.Y. 10016

inorganic and organic chemistry

AUGUSTINE ***Catalytic Hydrogenation***

(Techniques and Applications in Organic Synthesis Series)

by ROBERT L. AUGUSTINE, *Department of Chemistry, Seton Hall University, South Orange, New Jersey*

200 pages, illustrated. 1965

CONTENTS: Introduction • Apparatus and techniques • Catalysts and conditions • Hydrogenation of functional groups I • Hydrogenation of functional groups II • Hydrogenolysis.

AUGUSTINE ***Reduction***

(Techniques and Applications in Organic Synthesis Series)

edited by ROBERT L. AUGUSTINE, *Department of Chemistry, Seton Hall University, South Orange, New Jersey*

256 pages, illustrated. 1968

CONTENTS: Chemistry of mixed hydrides, *M. N. Rerick.* Dissolving metal reductions, *M. Smith.* Deoxygenation of carbonyl compounds, *W. Reusch.*

AUGUSTINE and TRECKER ***Oxidation***
In 2 Volumes

(Techniques and Applications in Organic Synthesis Series)

edited by ROBERT L. AUGUSTINE, *Department of Chemistry, Seton Hall University, South Orange, New Jersey,* and DAVID J. TRECKER, *Union Carbide Corporation, South Charleston, West Virginia.*

Vol. 1 edited by ROBERT L. AUGUSTINE
386 pages, illustrated. 1969

Vol. 2 edited by ROBERT L. AUGUSTINE and DAVID J. TRECKER
216 pages, illustrated. 1971

Provides concise and critical evaluations of important reactions in organic synthesis and makes available all aspects of oxidation. Serves as an easily accessible tool for the organic chemist.

CONTENTS:

Volume 1: Hydrocarbon oxidation using transition metal compounds, *D. G. Lee.* Oxidation of oxygen- and nitrogen-containing functional groups with transition metal compounds, *D. G. Lee.* Selenium dioxide oxidation, *E. N. Trachtenberg.* Glycol cleavage and related oxidations, *A. S. Perlin.* Peracid and peroxide oxidations, *S. N. Lewis.* Ozonization, *J. S. Belew.*

Volume 2: Sulfoxide-carbodiimide and related oxidations, *J. G. Moffatt.* Photosensitized oxygenations, *W. R. Adams.* Epoxidation of olefins by hydroperoxides, *R. Hiatt.* Metal ion-catalyzed oxidation of organic substrates with peroxides, *A. R. Doumaux, Jr.*

BAIZER ***Organic Electrochemistry: An Introduction and a Guide***

edited by MANUEL M. BAIZER, *Monsanto Company, St. Louis, Missouri*

1,096 pages, illustrated. 1973

Provides in one volume for the first time a comprehensive guide to organic electrochemistry, written by organic chemists who have made notable contributions to the field. Explores the capabilities and limitations of organic electrochemistry: its principles, methods, problems; its utility in synthesis and in elucidation of mechanisms of organic reactions; and its fertility as a research area. Major emphasis is on the detailed explanation of the chemical events which accompany electron–transfer steps. Of utmost interest to all organic chemists, as well as research workers involved in electrochemistry.

CONTENTS: Introductory survey, *M. Baizer.* Basic concepts, *G. Cauquis.* Methods for the elucidation of organic electrochemical reactions, *G. Cauquis and V. Parker.* Relation between micro and macro phenomena, *P. Zuman.* Practical problems in electrolysis, *H. Lund and P. Iversen.* Cathodic reactions of hydrocarbons, *R. Dietz.* Electrochemical reduction of organic halides, *M. Rifi.* Cathodic reduction of nitro compounds, *H. Lund.* Saturated carbonyl compounds and derivatives, *L. Feoktistov and H. Lund.* α,β-unsaturated carbonyls, *M. Baizer.* Carboxylic acids and derivatives, *L. Eberson.* Onium compounds, *L. Horner.* Hydrocarbons, *L. Eberson.* Carboxylic acids, *L. Eberson.* Anodic oxidation of amines, *V. Parker.* Anodic oxidation of oxygen-containing compounds, *V. Parker*. Anodic oxidation of sulfur-containing compounds, *V. Parker*. Electrolysis of heterocyclic compounds, *H. Lund*. Organometallic syntheses, *H. Lehmkuhl*. Electrolytic reductive coupling, *M. Baizer*. Oxidative coupling, *K. Nyberg*. Cleavages, *L. Horner and H. Lund*. Anodic substitution, *L. Eberson*. Amalgam reductions, *H. Lund*. Electrogenerated reagents, *R. Dietz and H. Lund*. Use of the solvated electron, *H. Lund*. Comparison between electrochemical and analogous chemical reactions, *L. Eberson*. Stereochemistry of organic electrode processes, *L. Eberson and L. Horner*. Industrial electroorganic chemistry, *D. Danly*. Electrochemical polymerization, *G. Parravano*. Organic electrocatalysis, *M. Baizer*.

CARLIN *Transition Metal Chemistry: A Series of Advances*

edited by RICHARD L. CARLIN, *Department of Chemistry, University of Illinois at Chicago Circle*

Vol. 1 320 pages, illustrated. 1965
Vol. 2 360 pages, illustrated. 1966
Vol. 3 376 pages, illustrated. 1966
Vol. 4 368 pages, illustrated. 1968
Vol. 5 320 pages, illustrated. 1969
Vol. 6 344 pages, illustrated. 1970
Vol. 7 352 pages, illustrated. 1972

A series devoted to the science of the transition elements, one of the oldest and largest areas of research. Geared to research scientists, graduate students, and teachers of advanced inorganic chemistry.

CONTENTS:

Volume 1: Electronic structure and stereochemistry of cobalt(II), *R. L. Carlin*. Aqueous chemistry of chromium(III), *J. E. Earley and R. D. Cannon*. Hydride complexes of transition metals, *A. P. Ginsberg*. Electronic structures of square planar metal complexes, *H. B. Gray*.

Volume 2: Reactions of ligands coordinated with transition metals, *J. P. Collman*. Transition metal ions as reagents in metallo-enzymes, *A. E. Dennard and R. J. P. Williams*. Optical activity in inorganic and organic compounds, *A. D. Liehr*.

Volume 3: Electronic structures of some organometallic molecules, *D. A. Brown*. Equilibria of complexes in nonaqueous solvents, *L. I. Katzin*. Electron spin resonance of transition-metal complexes, *B. R. McGarvey*. Fluorescent lifetimes of trivalent rare earths, *G. E. Peterson*. Conformations of coordinated chelates, *A. M. Sargeson*.

Volume 4: Spectra of Re^{4+} in cubic crystal fields, *P. B. Dorain*. Paramagnetic relaxation in solutions, *W. B. Lewis and L. O. Morgan*. Nature of transition between high-spin and low-spin octahedral complexes of transition metals, *R. L. Martin and A. H. White*. Electronic structure and stereochemistry of nickel (II), *L. Sacconi*. Coordination compounds of unsaturated 1,2-dithiols and 1,2-dithioketones, *G. N. Schrauzer*.

Volume 5: Spectra of chromium(III) complexes, *L. S. Forster*. Copper complexes, *W. E. Hatfield and R. Whyman*. Metal-metal exchange interactions, *G. F. Kokoszka and G. Gordon*.

Volume 6: Theory of bridge bonding and the structure of binuclear coordination compounds, *B. Jezowska-Trzebiatowska and W. Wojciechowski*. Amine complexes of chromium (III), *C. S. Garner and D. A. House*.

Volume 7: Magnetic phase transitions at low temperatures, *J. E. Rives*. Preparation and properties of high valent first-row transition metal oxides and halides, *C. Rosenblum and S. L. Holt*. Magnetic anisotropy, *S. Mitra*.

CHAPMAN *Organic Photochemistry*

a series edited by ORVILLE L. CHAPMAN, *Department of Chemistry, Iowa State University, Ames*

Vol. 1 352 pages, illustrated. 1967
Vol. 2 248 pages, illustrated. 1969
Vol. 3 320 pages, illustrated. 1973

Presents a critical summary of organic photochemistry drawing together seemingly unrelated facts, summarizing progress, and clarifying problems.

CONTENTS:

Volume 1: Photochemical transformations of cyclohexadienones and related compounds, *P. J. Kropp*. Photochemical transformations of small-ring carbonyl compounds, *A. Padwa*. Photo-Fries reaction and related arrangements, *V. I. Stenberg*. Photochemistry of troponoid compounds, *D. J. Pasto*. Photochemistry of olefins, *G. Fonken*. Photocyclization of stilbenes, *F. Stermitz*. Photocycloaddition reactions, *O. L. Chapman and G. Lenz*.

Volume 2: Photosensitization by energy transfer, *N. J. Turro, J. C. Dalton, and D. S. Weiss*. Photodimerizations, *D. J. Trecker*. Photochemistry of heteroaromatic nitrogen compounds, *P. Beak and W. R. Messer*. Photochemical additions to multiple bonds, *D. Elad*.

Volume 3: The Cis-Trans photoisomerization of olefins, *J. Saltiel, J. D'Agostino, E. Megarity, L. Metts, K. Neuberger, M. Wrighton, and O. Zafiriou*. Photochemistry of three-membered heterocycles, *N. Bertoniere and G. Griffin*. Photochemistry of cyclic ketones, *O. Chapman and D. Weiss*.

CHINN *Selection of Oxidants in Synthesis:* Oxidation at the Carbon Atom

(Oxidation in Organic Chemistry Series, Volume 1)

by LELAND J. CHINN, *G. Searle and Co., Chicago*

204 pages, illustrated. 1971 $13.50

(continued)

CHINN *(continued)*

Emphasizes stereoselectivity and the stereochemical aspects of the oxidative process. Useful to research chemists who are engaged in organic synthesis and graduate students and advanced undergraduate students in chemistry who are interested in the rationale behind the selection of an oxidizing agent for a particular transformation.

CONTENTS: Oxidation numbers • Formation of alcohols and ethers • Formation of alkyl hydroperoxides by autoxidation • Aldehydes and ketones from alcohols, halides, and amines • Formation of carboxylic acids from aldehydes • Generation of C=C or C–C bonds by dehydrogenation • Oxidation of olefinic double bonds • Oxidative cleavage of the C-C bond • Oxidation of alkyl groups to aldehydes • Oxidation of alcohols to acids and amines to amides • Oxidative cleavage of C–C and C=C bonds involving four electrons • Oxidation of methyl groups to carboxylic acids • Oxidative cleavage of C–C and C=C bonds involving six or more electrons.

COETZEE and RITCHIE *Solute-Solvent Interactions*

edited by J. F. COETZEE, *Department of Chemistry, University of Pittsburgh, Pennsylvania and* CALVIN D. RITCHIE, *Department of Chemistry, State University of New York, Buffalo*

672 pages, illustrated. 1969

Directed mainly to research workers in all areas of chemistry, the book contains chapters written by research workers who are eminently prominent in their respective fields. Treats solvent effects in far greater depth than does any book published thus far.

CONTENTS: Heats and entropies of ionization, *J. W. Larson and L. G. Hepler.* Medium effects and pH in nonaqueous solvents, *R. G. Bates.* Acidity functions, *R. H. Boyd.* Interactions in dipolar aprotic solvents, *C. D. Ritchie.* Selective solvation of ions in mixed solvents, *H. Schneider.* Solvent isotope effect on thermodynamics of nonreacting solutes, *E. M. Arnett and D. R. McKelvey.* Solvent isotope effects for equilibria and reactions, *P. M. Laughton and R. E. Robertson.* Organoalkali compounds in ethers, *J. F. Garst.*

COOK *Enamines: Synthesis, Structure, and Reactions*

edited by A. GILBERT COOK, *Department of Chemistry, Valparaiso University, Indiana*

528 pages, illustrated. 1969

Reviews and correlates in depth the synthetic, mechanistic, and physical properties of enamines.

CONTENTS: Structure and physical properties of enamines, *S. K. Malhotra.* Methods and mechanisms of enamine formation, *L. W. Haynes.* Hydrolysis of enamines, *E. J. Stamhuis.* Electrophilic substitutions and additions to enamines, *G. H. Alt.* Nucleophilic addition to iminium salts, *J. V. Paukstelis.* Cycloaddition reactions of enamines, *A. G. Cook.* Heterocyclic enamines, *O. Červinka.* Enamines in organic synthesis, *M. E. Kuehne.*

CRAGG *Organoboranes in Organic Synthesis*

(Studies in Organic Chemistry Series, Volume 1)

by GORDON CRAGG, *Department of Chemistry, University of Cape Town, Rondebosch, South Africa*

432 pages, illustrated. 1973

Illustrates the potential usefulness in organic synthesis of new methods for the synthesis of compounds containing a wide variety of functional groups, and examines in detail the preparation of organoboranes and their applications to the synthesis of the various classes of organic compounds. Of great value to chemists involved in synthetic, organic and organometallic research, as well as to teachers of organic chemistry.

CONTENTS: Organoboranes in organic synthesis • Synthesis of organoboranes • Hydroboration of alkenes • Oxidation of organoboranes • Hydroboration of functionalized alkenes • Hydroboration of dienes and polyenes • Hydroboration of alkynes • Synthesis of carbon chains and rings • Synthesis of functional derivatives via organoboranes • Reduction of functional groups.

DALEY and O'MALLEY *Problems in Chemistry*

by HENRY O. DALEY, *Department of Chemistry, Bridgewater State College, Bridgewater, Massachusetts,* and ROBERT F. O'MALLEY, *Department of Chemistry, Boston College, Massachusetts*

in preparation. 1974

Provides the beginning chemistry student with a large number of problems which illustrate chemical principles. The problems are based, wherever possible, on experimental results obtained by chemists, and the references are cited.

DASENT *Nonexistent Compounds: Compounds of Low Stability*

by W. E. DASENT, *Department of Chemistry, Victoria University of Wellington, New Zealand*

192 pages, illustrated. 1965

CONTENTS: General considerations • Compounds whose instability is a consequence of the restriction of first-row atoms (Li to F) to valence octet • Compounds whose instability is related to reluctance of certain atoms to undergo self-linkage or catenation • Compounds whose instability is a consequence of reluctance of atoms of second (and subsequent) rows to form multiple bonds involving $P\pi$-$P\pi$ overlap • Compounds whose instability has been attributed to operation of an "inert pair effect" • Compounds of elements of the first long period whose highest oxidation state is relatively unstable • Some compounds of the noble gases • Miscellaneous.

DENNEY *Techniques and Methods of Organic and Organometallic Chemistry*

edited by DONALD B. DENNEY, *School of Chemistry, Rutgers—The State University, New Brunswick, New Jersey*

248 pages, illustrated. 1969

Presents critical reviews of methods and techniques used in the laboratory. Is of interest to organic and organo-metallic research workers—organic and inorganic chemists—investigating the study of organic compounds.

CONTENTS: Ozonolysis of organic compounds, *R. W. Murray*. High-vacuum techniques, *A. C. Bond*. Applications of fused salts in organic chemistry, *J. E. Gordon*. Nonaqueous titrations for organic analysis, *C. A. Streuli*.

FLURRY *Molecular Orbital Theories of Bonding in Organic Molecules*

(Applied Quantum Chemistry Series, Volume 1)

by ROBERT L. FLURRY, JR., *Department of Chemistry, Louisiana State University, New Orleans*

244 pages, illustrated. 1968

CONTENTS: Introduction • Free electron theory • Simple LCAO theory • Perturbation methods • Solving LCAO equations • Group theory and symmetry orbitals • Atomic theory • Self-consistent molecular orbital methods • Sigma bonds • Applying the theories.

FONKEN and JOHNSON *Chemical Oxidations with Microorganisms*

(Oxidation in Organic Chemistry Series, Volume 2)

by GUNTHER S. FONKEN and ROY A. JOHNSON, *The Upjohn Company, Kalamazoo, Michigan*

320 pages, illustrated. 1972

Presents organic chemists with an introduction to the many microbial oxidations of organic molecules that can be considered for use in the laboratory. Examines all major oxidative categories and several minor ones in a thorough manner. A highly useful work for organic chemists; undergraduate and graduate students in organic chemistry; industrial microbiologists and graduate students in microbiology; biochemists; and medicinal chemists working in the area of drug metabolism.

CONTENTS: Nonactivated carbon hydroxylation • Allylic oxygenations • Olefinic oxygenation • Aromatic ring hydroxylation • Aromatic ring opening • Microbiological Baeyer-Villiger oxidation • β-oxidation • Alkyl dehydrogenation • Alcohol dehydrogenation • Oxidation of amino to nitro • Oxidation of amino to hydroxyl • Sulfur oxygenation • Other microbial oxidations • Practical experimental methods.

HUYSER *Methods in Free-Radical Chemistry*

a series edited by EARL S. HUYSER, *Department of Chemistry, The University of Kansas, Lawrence*

Vol. 1 224 pages, illustrated. 1969

Vol. 2 256 pages, illustrated. 1969

Vol. 3 184 pages, illustrated. 1972

Vol. 4 in preparation. 1974

Written by authorities in various areas of free radical reactions, these volumes are of primary interest to the organic chemist.

CONTENTS:

Volume 1: Free-radical study by electron paramagnetic resonance, *L. Kevan*. Free radicals and photochemical reactions, *D. Neckers*. Free-radical chlorination of organic molecules, *M. L. Poutsma*.

Volume 2: Thiyl radicals, *R. M. Kellogg*. Free-radical brominations, *W. Thaler*.

Volume 3: Electrolytically generated radicals, *H.-G. Gilde*. One-electron reduction of aromatic cations, *M. Siskin*. Free-radical chlorinations via nitrogen cation radicals, *N. C. Deno*.

JACOBSON and CROSBY *Naturally Occurring Insecticides*

edited by MARTIN JACOBSON, *U. S. Department of Agriculture, Beltsville, Maryland* and DONALD G. CROSBY, *University of California, Davis*

600 pages, illustrated. 1971

A critical review on natural insecticides prompted by growing public alarm over the dangerous side effects of conventional pesticides and the prospects of many harmful

(continued)

JACOBSON and CROSBY *(continued)*

insects becoming resistant to the synthetic insecticides. Of value to entomologists, organic, analytical, and agricultural chemists, insect physiologists, and botanists.

CONTENTS: Pyrethroids, *M. Matsui and I. Yamamoto.* Rotenone and the rotenoids, *H. Fukami and M. Nakajima.* Nicotine and other tobacco alkaloids, *I. Schmeltz.* The unsaturated isobutylamides, *M. Jacobson.* Minor insecticides of plant origin, *D. Crosby.* Anthropod venoms as insecticides, *R. Beard.* Ant secretions and cantharidin, *G. Cavill and D. Clark.* Juvenile hormones, *W. Bowers.* The ecdysones, *D. Horn.* Bacillus thuringiensis as a microbial insecticide, *T. Angus.* Destruxins and piericidins, *S. Tamura and N. Takahashi.*

JONES *Inorganic Vibrational Spectroscopy*

by LLEWELLYN H. JONES, *Los Alamos Scientific Laboratory, University of California, Los Alamos, New Mexico*

Vol. 1 232 pages, illustrated. 1971

Presents a discussion of the use of vibrational spectroscopy to determine the strengths of chemical bonds, the bonding directional forces, and the interactions of one bond or bond angle with the others in a molecule. A great help to the inorganic chemist who is interested in the applications of vibrational spectroscopy or in evaluating other workers' results and interpretations.

CONTENTS: Force constants: Their interpretation and determination • Potential constants of some diatomic and triatomic molecules from observed data • Force constants of halide complexes • Metal cyanide complexes • Metal carbonyls.

KAPLAN *Bridged Free Radicals*

by LEONARD KAPLAN, *University of Chicago, Illinois*

504 pages, illustrated. 1972

Presents a comprehensive and critical examination of bridging in free radicals. Organized according to the structure of bridged free radicals in order to simplify finding specific information. Of particular importance to all graduate students and research scientists in organic and physical chemistry.

CONTENTS: Definitions • Silyl-substituted radicals • Germyl-substituted radicals • Stannyl-substituted radicals • Chloro-substituted radicals • Bromo-substituted radicals • Iodo-substituted radicals • Sulfur-substituted radicals • Cyano-substituted radicals • Vinyl-substituted radicals.

LAGOWSKI *Modern Inorganic Chemistry*

by JOSEPH J. LAGOWSKI, *University of Texas, Austin*

832 pages, illustrated. 1973

The most important textbook in the field to be published in the last decade.

Designed to give advanced undergraduate or first-year graduate students a working knowledge of the fundamentals of modern inorganic chemistry in those particular areas classically associated with the field. Acknowledges the close relationship between experiment and theory but separates these two aspects of chemistry in order to demonstrate that unity of experimental results is possible without the use of extensive theoretical arguments.

Modern Inorganic Chemistry is an invaluable teaching tool for all professors of chemistry.

CONTENTS: Atoms and their characteristics • Ionic compounds • Covalent compounds: Experimental observations • Covalent compounds: Theories of bonding • Representative elements: A brief survey • Hydrogen • Solvent properties of covalent hydrides • Alkali metals • Alkaline-earth metals • Group III Representative Elements • Group IV elements • Group V elements • Group VI elements • The halogens • The rare gases • The transition elements • Properties of transition-metal coordination compounds • Transition-metal complexes: Theoretical aspects • Organic derivates of the transition metals.

LAWLESS and SMITH *Inorganic High-Energy Oxidizers: Synthesis, Structures, and Properties*

by EDWARD W. LAWLESS and IVAN C. SMITH, *Midwest Research Institute, Kansas City, Missouri*

320 pages, illustrated. 1968

CONTENTS: Introduction • General characteristics of advanced oxidizers • Fluoronitrogen compounds • Fluorine and fluorohalogen compounds • Oxygen compounds.

LEDNICER *Contraception: The Chemical Control of Fertility*

edited by DANIEL LEDNICER, *The Upjohn Company, Kalamazoo, Michigan*

280 pages, illustrated. 1969

Provides the interested student with a critical review of the antifertility research and serves to acquaint the biologist and the chemist with the literature and methods in the field.

CONTENTS: The reproductive cycle in the female, *R. B. Jaffe, G. Peréz-Palacios, and G. B. Serra.* The biology of steroidal contraceptives, *R. A. Edgren.* The chemistry of steroidal contraceptives, *P. D. Klimstra and F. B. Colton.* The biology of nonsteroidal antifertility agents, *L. Lerner.* The chemistry of nonsteroidal antifertility agents, *D. Lednicer.* Assays and screens in antifertility research, *R. L. Elton.*

MacDIARMID *The Bond to Carbon*

In 2 Parts

(Organometallic Compounds of the Group IV Elements, Volume 1)

edited by ALAN G. MACDIARMID, *Department of Chemistry, University of Pennsylvania, Philadelphia*

Part I 624 pages, illustrated. 1968

Part II 280 pages, illustrated. 1968

CONTENTS:

Part I: Physical basis of the chemistry of the Group IV elements, *E. Ebsworth.* Synthesis and reactions of the silicon-carbon bond, *C. Eaborn and R. Bott.*

Part II: Synthesis and properties of the germanium-carbon bond, *F. Glockling and K. Hooton.* Synthesis and properties of the tin-carbon bond, *J. Luijten and G. van der Kerk.* Synthesis and properties of the lead-carbon bond, *L. Willemsens and G. van der Kerk.*

MacDIARMID *The Bond to Halogens and Halogenoids*

In 2 Parts

(Oranometallic Compounds of the Group IV Elements, Volume 2)

edited by ALAN G. MACDIARMID, *Department of Chemistry, University of Pennsylvania, Philadelphia*

Part I 392 pages, illustrated. 1972

Part II 248 pages, illustrated. 1972

Describes methods of synthesis of the group IV element-halogen bond and the group IV element-halogenoid bond, examines the chemical reactions in which these bonds are cleaved, and discusses the important physical and spectroscopic properties related to the presence of these bonds in a molecule. Valuable reading for research chemists and senior students of organometallic, organic, and inorganic chemistry.

CONTENTS:

Part I: Synthesis and properties of the silicon-halogen and silicon-halogenoid bond, *C. Van Dyke.*

Part II: Synthesis and properties of the germanium-halogen and germanium-halogenoid bond, *J. Zuckerman.* Synthesis and properties of the tin-halogen and tin-halogenoid bond, *H. Clark and R. Puddephatt.* Synthesis and properties of the lead-halogen and lead-halogenoid bond, *S. Cook, F. Frey, and H. Shapiro.*

MANHAS and BOSE *Synthesis of Penicillin, Cephalosporin C, and Analogs*

by MAGHAR S. MANHAS and AJAY K. BOSE, *Stevens Institute of Technology, Hoboken, New Jersey*

136 pages, illustrated. 1969

Begins with a brief history of penicillin synthesis, and continues with a detailed description of the spectral methods dealing with penam, cepham, and related systems which are prone to various rearrangements. Concludes with a discussion on the synthesis leading to penicillin, cephalosporin C, and their analogs.

CONTENTS: Introduction • Physical properties • Synthesis of penicillin and analogs • Synthesis of cephalosporin C and analogs.

MARCH *Problems in Advanced Organic Chemistry*

by JERRY MARCH, *Adelphi University, Garden City, New York*

432 pages, illustrated. 1971

Consists of over 800 problems, most of which are actual examples taken from recent literature, with references given for each. Answers to about one third of the problems are found at the end of the book; literature references are given for the remainder. Can accompany any textbook used for advanced courses in organic chemistry, or may stand alone when no text is available.

CONTENTS: **Problem Sets:** Bonding • Stereochemistry • Acidity, mechanisms, and reactivity • Aliphatic nucleophilic substitution • Aromatic electrophilic substitution • Aliphatic electrophilic substitution • Aromatic nucleophilic substitution • Free-radical substitution • Addition to carbon-carbon multiple bonds • Addition to carbon-hetero multiple bonds • Eliminations • Rearrangements • Oxidations and reductions • Additional problems in synthesis and mechanisms • Nomenclature • Literature • **Answers.**

MARINSKY *Ion Exchange: A Series of Advances*

edited by JACOB A. MARINSKY, *Department of Chemistry, State University of New York, Buffalo*

Vol. 1 440 pages, illustrated. 1966

Vol. 2 256 pages, illustrated. 1969

Vol. 3 see MARINSKY and MARCUS for continuing volumes

". . . a brilliant work . . . The importance of this series lies in the vitality of its approach to the subject, together with the care with which hundreds of references are not simply collected, but are correlated, evaluated, and compared . . . throughout the books, a re^freshing originality of approach is found . . ."—T. V. Arden, *Chemistry and Industry*

CONTENTS:

Volume 1: Transport processes in membranes, *S. R. Caplan and D. C. Mikulecky.* Ion-exchange kinetics, *F. Helfferich.* Ion-exchange studies of complex formation, *Y. Marcus.* Liquid ion exchangers, *E. Högfeldt.* Precise studies of ion-exchange systems using microscopy, *D. H. Freeman.* Heterogeneity and physical chemical properties of ion-exchange resins, *L. S. Goldring.* Ion-exchange selectivity, *D. Reichenberg.* Resin selectivity in dilute to concentrated aqueous solutions, *R. M. Diamond and D. C. Whitney.* Interpretation of ion-exchange phenomena, *J. A. Marinsky.*

Volume 2: Ion exchange in glasses, *R. H. Doremus.* Ion exchange in molten systems, *E. C. Freiling and M. H. Rowell.* Ion-exchange properties of zeolites, *H. S. Sherry.* Interactions between organic ions and ion-exchange resins, *J. Feitelson.* Partition chromatography of sugars, sugar alcohols, and sugar derivatives, *O. Samuelson.* Synthesis of ion-exchange resins, *R. M. Wheaton and M. J. Hatch.*

MARINSKY and MARCUS *Ion Exchange and Solvent Extraction: A Series of Advances*

a series edited by JACOB A. MARINSKY, *Department of Chemistry, State University of New York, Buffalo,* and YIZHAK MARCUS, *Department of Inorganic and Analytical Chemistry, The Hebrew University, Jerusalem, Israel*

for earlier volumes see MARINSKY

Vol. 3 168 pages, illustrated. 1973

Vol. 4 264 pages, illustrated. 1973

Vol. 5 256 pages, illustrated. 1973

Vol. 6 in preparation. 1974

CONTENTS:

Volume 3: Extraction of metals by carboxylic acids, *D. Flett and M. Jaycock.* Solvent extraction with sulfonic acids, *G. Markovits and G. Choppin.* Nuclear magnetic resonance studies of organophosphorus extractants, *W. Stewart and T. Siddall, III.* Experience with the AKUFVE solvent extraction equipment, *J. Rydberg, H. Reinhardt, and J. Liljenzin.*

Volume 4: Ion exchange in nonaqueous and mixed solvents, *Y. Marcus.* Liquid exchange chromatography, *H. Walton.* Liquid ion exchange technology, *R. Kunin.* Electronic and ionic exchange properties, conductivity, and permselectivity of organic semiconductors and redox exchangers, *R. Buvet, M. Guillou, and L. Yu.* Equations for the evaluation of formation constants of complexed ion species in crosslinked and linear polyelectrolyte systems, *J. Marinsky.*

Volume 5: New inorganic ion exchangers, *A. Clearfield, G. Nancollas, and R. Blessing.* Application of ion exchange to element separation and analysis, *F. Strelow.* Pellicular ion exchange resins in chromatography, *C. Horvath.*

Volume 6: Carbon fiber structure, surface and surface treatments, *J. Larsen and T. Smith.* Isolation of drugs and related organic compounds by ion-pair extraction, *G. Schill.* The dynamics of liquid-liquid extraction processes, *G. Pollock and A. Johnson.* The application of the solubility parameter concept in liquid-liquid extraction, *H. Irving.*

MUETTERTIES *Transition Metal Hydrides*

(The Hydrogen Series, Volume 1)

edited by EARL L. MUETTERTIES, *E. I. du Pont de Nemours & Company, Wilmington, Delaware*

360 pages, illustrated. 1971

The first book to answer the need for a systematic and definitive presentation of the chemistry of molecular transition metal hydrides. Of great value to inorganic and physical chemists, as well as to organic chemists seeking stereospecific syntheses.

CONTENTS: Physical properties of hydrogen, deuterium, and tritium, *E. L. Muetterties.* The transition metal-hydrogen interaction, *E. L. Muetterties.* Molecular structures of transition metal hydride complexes, *B. A. Frenz and J. A. Ibers.* Stereochemistry and stereochemical nonrigidity in transition metal hydrides, *J. P. Jesson.* Systematics of transition metal hydrides, *R. A. Schunn.* Role of transition metal hydrides in homogeneous catalysis, *C. A. Tolman.*

MUETTERTIES and KNOTH *Polyhedral Boranes*

by EARL L. MUETTERTIES and WALTER H. KNOTH, *Central Research Department, Experimental Station, E. I. duPont de Nemours & Co., Wilmington, Delaware*

208 pages, illustrated. 1968

CONTENTS: Introduction • Structure and theory • Polyhedral rearrangements • Structure-reactivity correlation • Synthesis of polyhedral boranes • Chemistry • Nomenclature and numbering conventions • Recent developments.

OWEN *Characterization of Organic Compounds by Chemical Methods: An Introductory Laboratory Textbook*

by T. C. OWEN, *University of South Florida, Tampa*

256 pages, illustrated. 1969

CONTENTS: Introduction • Functional group organic chemistry • Equipment and techniques • Functional group identification • Selection and preparation of solid derivatives • Tables of physical properties.

RAO *Xanthates and Related Compounds*

by S. R. RAO, *Indian Institute of Technology, Delhi, India*

512 pages, illustrated. 1971

A comprehensive survey of all aspects of the basic chemistry of xanthates and the related compounds, dixanthogens. Directed to mineral flotation engineers and investigators, textile chemists, chemical analysts, and investigators in organo-sulfur chemistry.

CONTENTS: Introduction • Methods of preparation and purification • General chemical properties and reactions • Methods of analysis of xanthates • Dixanthogens • Structural chemistry of xanthates and dixanthogens • Xanthates in mineral flotation • Analytical chemical applications • Xanthates in the cellulose industry • Miscellaneous applications of xanthates and dixanthogens.

RIFI and COVITZ *Introduction to Organic Electrochemistry*

(Techniques and Applications of Organic Synthesis Series)

by M. R. RIFI and F. H. COVITZ, *Union Carbide Corporation, Bound Brook, New Jersey*

424 pages, illustrated. 1973

Provides an up-to-date introduction to the field of organic electrochemistry. Discusses theoretical aspects and emphasizes the practical advantages of many experimental procedures and specific types of synthesis and cell designs. Primarily intended for organic chemists in academic or industrial environments who need the basic information to initiate a program in this field.

CONTENTS: Basic principles • Apparatus and techniques • Reduction of functional groups • Oxidation of functional groups • Electro-initiated polymerization • Electrocoating.

SAWYER *Organotin Compounds*

In 3 Volumes

edited by ALBERT SAWYER, *Department of Chemistry, University of New Hampshire, Durham*

Vol. 1 272 pages, illustrated. 1971

Vol. 2 389 pages, illustrated. 1971

Vol. 3 384 pages, illustrated. 1973

Provides contemporary, comprehensive coverage of the entire field of organotin chemistry, prepared by individuals who are well informed in their specialized areas. Of particular interest to organic, inorganic, polymer, surface, and physical chemists working in the field of organometallics and organotin chemistry.

CONTENTS:

Volume 1: Introduction, *G. van der Kerk and J. Luijten.* Organotin hydrides, *E. Kupchik.* Organotin halides, *G. van der Kelen, E. van den Berghe, and L. Verdonck.* Organotin compounds with Sn-O bonds: Organotin alkoxides, oxides, and related compounds, *A. Bloodworth and A. Davies.*

Volume 2: Organotin compounds with Sn-O bonds. Organotin carboxylates, salts, and complexes, *R. Okawara and M. Ohara.* Organotin compounds with Sn-S, Sn-Se, and Sn-Te bonds, *H. Schumann, I. Schumann-Ruidisch, and M. Schmidt.* Organotin compounds with Sn-N bonds, *K. Jones and M. F. Lappert.* Organotin compounds with Sn-P, Sn-As, Sn-Sb, and Sn-Bi bonds, *H. Schumann, I. Schumann-Ruidisch, and M. Schmidt.*

Volume 3: Organotin compounds with Sn-C bonds without Sn-Sn bonds, *M. Gielen and J. Nasielski.* Organotin compounds with Sn-Sn bonds, *A. K. Sawyer.* Organotin compounds with Sn-other metal bonds, *M. J. Newlands.* Applications and biological effects of organotin compounds, *J. G. A. Luijten.* Organotin polymers, *M. C. Henry and W. E. Davidsohn.* Analysis of organotin compounds, *C. R. Dillard.*

SCHRAUZER *Transition Metals in Homogeneous Catalysis*

edited by G. N. SCHRAUZER, *University of California, La Jolla*

428 pages, illustrated. 1972

The first book devoted entirely to homogeneous catalysis by transition metal complexes. Bridges the gap between academic teaching and industrial research in homogeneous catalysis.

(continued)

SCHRAUZER *(continued)*

CONTENTS: Catalysis: Fundamental aspects and scope, *G. Schrauzer*. Hydrogenation and dehydrogenation, *J. Kwiatek*. π-Allyl system in catalysis, *W. Keim*. Homogeneous metal catalyzed oxidation of organic compounds, *E. Stern*. Carbonylation, *D. Thompson and R. Whyman*. Catalysis of symmetry forbidden reactions, *F. Mango and J. Schachtschneider*. Electron-transfer catalysis, *R. Linck*.

SENNING *Sulfur in Organic and Inorganic Chemistry*

In 3 Volumes

edited by ALEXANDER SENNING, *Aarhus University, Denmark*

Vol. 1 400 pages, illustrated. 1971

Vol. 2 376 pages, illustrated. 1972

Vol. 3 480 pages, illustrated. 1972

Presents an integrated treatment of sulfur chemistry across the traditional lines of inorganic, organic, physical-chemical, and biochemical aspects of this field, and gives a critical and authoritative account of the principles found useful in the systemization of known facts and in the design of new research. Of prime importance to academic researchers in inorganic, organic, analytical, and physical chemistry, and biochemistry. Also of interest to industrial researchers working with drugs or pesticides and to graduate students.

CONTENTS:

Volume 1: Introduction, *A. Senning*. The sulfur-silicon bond, *A. Haas*. The sulfur-nitrogen bond, *H. W. Roesky*. The sulfur-phosphorus bond, *L. Almasi*. The sulfur-oxygen bond, *H. H. Szmant*. The sulfur-sulfur bond, *J. L. Kice*. The sulfur-fluorine bond, *S. P. von Halasz and O. Glemser*. The sulfur-chlorine bond, *C. R. Russ and I. B. Douglass*. The sulfur-bromine bond, *P. S. Magee*. The sulfur-iodine bond, *J. P. Danehy*.

Volume 2: The chemistry of atomic sulfur, *O. P. Strausz*. Diatomic species containing sulfur, *B. Meyer, D. Jensen, and T. Oommen*. Bond energy terms in the chemistry of sulfur, selenium, and tellurium, *D. A. Johnson*. Oxyacids of sulfur, *M. Schmidt*. Metabolic pathways of organic sulfur compounds, *G. A. Maw*. The pharmacology and toxicology of inorganic sulfur compounds, *B. Sörbo*. The mass spectra of sulfur compounds, *G. Schroll and S.-O. Lawesson*. Mixed sulfur halides, *F. Seel*. Commercially important sulfur compounds, *R. Leclercq*. Chromatographic techniques in sulfur chemistry, *E. R. Cole and R. F. Bayfield*.

Volume 3: Reactions of elemental sulfur with inorganic, organic, and metal-organic compounds, *H. Schumann*. Inorganic and organic polysulfides, *T. Pickering and A. V. Tobolsky*. The quantum chemistry of sulfur compounds, *J. Fabian*. Steric aspects of sulfur chemistry, *P. Laur*. NMR spectra of sulfur compounds, *C. Brown*. Labeled sulfur compounds, *E. Blasius, W. Neumann, and H. Wagner*. Thionethiol tautomerism, *R. Mayer*. The nomenclature of sulfur compounds and their selenium and tellerium analogs, *K. L. Loening*. Nucleophilicity of organic sulfur compounds, *M. J. Janssen*.

TARRANT *Fluorine Chemistry Reviews*

a series edited by PAUL TARRANT, *Department of Chemistry, University of Florida, Gainesville*

Vol. 1 430 pages, illustrated. 1967

Vol. 2 192 pages, illustrated. 1968

Vol. 3 160 pages, illustrated. 1969

Vol. 4 216 pages, illustrated. 1969

Vol. 5 200 pages, illustrated. 1971

Vol. 6 168 pages, illustrated. 1973

A series concerned with the theoretical and practical aspects of fluorine chemistry which provides complete coverage of current research in this rapidly expanding field. A valuable asset to all researchers involved in the field of fluorine chemistry.

CONTENTS:

Volume 1: Synthesis, compounding, and properties of nitroso rubbers, *M. C. Henry, C. G. Griffis, and E. C. Stump*. Electrochemical fluorination, *S. Nagase*. Fluoroketenes, *Yu. A. Cheburkov and I. L. Knunyants*. Hexafluoroacetone, *C. G. Krespan and W. J. Middleton*. Fluorocarbon toxicity and biological action, *J. W. Clayton, Jr*. Diels-Alder reactions of organic fluorine compounds, *D. R. A. Perry*. Methods for the introduction of hydrogen into fluorinated compounds, *F. J. Mettille and D. J. Burton*. Reactions involving fluoride ion and polyfluoroalkyl anions, *J. A. Young*.

Volume 2: Cycloaddition reaction of fluoroolefins, *W. H. Sharkey*. Reaction of halogenated cycloalkenes with nucleophiles, *J. D. Park, R. J. McMurtry, and J. H. Adams*. Ionization potentials and molecule-ion dissociation energies for diatomic metal halides, *J. W. Hastie and J. L. Margrave*. NMR spectra of nitrogen-fluorine compounds, *W. S. Brey, Jr. and J. B. Hynes*. ^{19}F chemical shifts and coupling constants of fluoroxy compounds, *C. J. Hoffman*.

Volume 3: Fluorine compounds in anesthesiology, *E. Larsen*. Reactions of fluoroolefins with electrophilic reagents, *B. Dyatkin, E. Mochalina, and I. Knunyants*. Fluoroalicyclic derivatives of metals and metalloids, *W. Cullen*. Phosphorus, arsenic, and antimony pentafluorophenyl compounds, *M. Fild and O. Glemser*.

Volume 4: Polyhaloalkyl derivatives of sulphur, *R. Dresdner and T. Hooper*. The chemistry of fluorinated acetylenes, *M. Bruce and W. Cullen*. The chemistry of aliphatic fluoronitrocarbons, *E. Bissell*.

Volume 5: Electron spin resonance of irradiated organic fluorine compounds, *M. Iwasaki*. The preparation and chemistry of fluorinated diazo compounds and diazirines, *C. G. Krespan and W. J. Middleton.* Fluorine-containing epoxides, *P. Tarrant, C. G. Allison, K. P. Barthold, and E. C. Stump, Jr.* The chemistry of fluoroalkyl radicals, *A. P. Stefani.*

Volume 6: The radiation chemistry of polyfluorinated organic compounds, *H. Sutcliffe and I McAlpine.* Fluoro-ß-diketones and metal fluoro-ß-diketonates, *P. Mushak, M. Glenn, and J. Savory.* NMR spectra and characteristic frequencies of compounds containing N-S-F bonds, *H. Horn.*

TAYLOR and BATTERSBY *Cyclopentanoid Terpene Derivatives*

(Organic Substances of Natural Origin Series, Volume 2)

edited by W. I. TAYLOR, *International Flavors and Fragrances, Inc., Union Beach, New Jersey,* and A. R. BATTERSBY, *University Chemical Laboratory, University of Cambridge, England*

448 pages, illustrated. 1969

CONTENTS: The iridoid glycosides and similar substances, *J. M. Bobbitt and K.-P. Segebarth.* Picrotoxin, *C. J. Coscia.* Insect terpenoids and nepetalactone, *G. W. K. Cavill.* Monoterpene alkaloids, *W. C. Wildman, J. Le Men, and (in part) K. Wiesner.* Lower furanoterpenes, *T. Kubota.* Sesquiterpene alkaloids, *O. E. Edwards.*

TAYLOR and BATTERSBY *Oxidative Coupling of Phenols*

(Organic Substances of Natural Origin Series, Volume 1)

edited by W. I. TAYLOR, *International Flavors and Fragrances, Inc., Union Beach, New Jersey,* and A. R. BATTERSBY, *University Chemical Laboratory, University of Cambridge, England*

408 pages, illustrated. 1967

CONTENTS: Phenol coupling, *H. Musso.* Some natural products derived by phenol oxidation, *A. I. Scott.* Phenol oxidations in the alkaloid field, *A. R. Battersby.* Biochemical aspects of oxidative coupling of phenols, *B. R. Brown.* Aphid pigments, *D. W. Cameron and Lord Todd.* Lignin – a natural polymeric product of phenol oxidation, *J. M. Harkin.* Lignans and cyclolignans, *K. Weinges and R. Spänig.*

TAYLOR and FARNSWORTH *Vinca Alkaloids*

edited by WILLIAM I. TAYLOR, *International Flavors and Fragrances, Union Beach, New Jersey* and NORMAN R. FARNSWORTH, *Department of Pharmacognosy and Pharmacology, University of Illinois, Chicago*

in preparation. 1974

Brings together in a single volume all the important aspects of *vinca* alkaloids. An excellent text for graduate students who are well grounded in organic chemistry, botany, or pharmacology, and a useful reference for established researchers.

CONTENTS: A synopsis of the genus *vinca* including its taxonomic and nomenclatural history, *W. Stearn.* The phytochemistry of *vinca* species, *N. Farnsworth.* The chemistry of the *vinca* alkaloids, *W. Taylor.* Chemotaxonomy of *vinca* species, *J. Trojánek, M. Novácek, and F. Stary.* The commercial cultivation of *vinca minor, K. Szász and G. Mark.* The pharmacology of *vinca* species and their alkaloids, *M. Hava.*

VIEHE *Chemistry of Acetylenes*

edited by H. G. VIEHE, *Union Carbide European Research Associates, Brussels, Belgium*

1320 pages, illustrated. 1969

Presents for the first time in many years an up-to-date coverage of the chemistry of acetylene compounds. The material has been arranged in such an order as to develop and cover as many aspects of the field as possible. Of significant value to the entire chemical industry.

CONTENTS: **Part I: Theoretical:** Structure and physical properties of acetylenic compounds; the nature of the triple bond, *J. Dale.* **Part II: Syntheses:** Synthesis of acetylenes and polyacetylenes by elimination reactions, *G. Köbrich and P. Buck.* Synthesis of acetylenes and polyacetylenes by substitution reactions, *W. Ziegenbein.* **Part III: Reactions:** Ionic additions to acetylenes, *E. Winterfeldt.* Free radical additions to acetylenes, *M. Julia.* Partial catalytic hydrogenation of acetylenes, *H. Gutmann and H. Lindlar.* Propargylic rearrangements, *J. H. Wotiz.* Cyclic compounds from acetylenes, *R. Fuks and H. G. Viehe.* Coupling of acetylenes, *P. Cadiot and W. Chodkiewicz.* **Part IV: The Chemistry of Heterosubstituted Acetylenes:** 1-Halogenoacetylenes, *S. Delavarenne and H. G. Viehe.* Acetylenic ethers and thioethers, *L. Brandsma, H. J. T. Bos, and J. F. Arens.* Ynamines, *H. G. Viehe.* Acetylenic derivatives of groups IIIb, IVb, and Vb, *P. Cadiot and W. Chodkiewicz.* **Part V: Chemistry of Special Acetylenes:** Acetylenes from nature, *F. Bohlmann.* Cyclic acetylenes, *A. Krebs.* Dehydrobenzene, *R. W. Hoffmann.* Dehydrohetarenes, *H. J. den Hertog and H. C. van der Plas.*

OTHER BOOKS OF INTEREST

BOLKER ***Natural and Synthetic Polymers:*** *An Introduction*

by HENRY I. BOLKER, *Department of Chemistry, McGill University, Montreal, Quebec*

in preparation. 1973

DURIG ***Vibrational Spectra and Structure:*** *A Series of Advances*

a series edited by JAMES R. DURIG, *University of South Carolina, Columbia*

Vol. 1 224 pages, illustrated. 1972

Vol. 2 in preparation. 1973

MATAGA and KUBOTA ***Molecular Interactions and Electronic Spectra***

by NOBORU MATAGA, *Osaka University, Osaka, Japan* and TANEKAZU KUBOTA, *Shionogi Research Laboratory, Osaka, Japan*

520 pages, illustrated. 1970

MUROV ***Handbook of Photochemistry***

by STEVEN L. MUROV, *Department of Chemistry, State University of New York at Stony Brook*

in preparation. 1973

VOGL and FURUKAWA
Polymerization of Heterocyclics

edited by OTTO VOGL, *Department of Polymer Science and Engineering, University of Massachusetts, Amherst,* and JUNJI FURUKAWA, *Department of Synthetic Chemistry, Kyoto University, Japan*

216 pages, illustrated. 1973

JOURNALS OF INTEREST

MOLECULAR PHOTOCHEMISTRY

editor: DONALD VALENTINE, JR., *Hoffman-LaRoche, Inc., Nutley, New Jersey*

A periodical designed to encourage further discourse between photochemists and spectroscopists so that the symbiotic relationship between the two fields may be fruitfully broadened. Mechanisms of photochemical reactions, primary processes, structure-reactivity relationships, spectral-reactivity relationships, structures of molecules in excited states, radiationless processes, and electronic energy transfer are discussed.

4 issues per volume

SYNTHESIS IN INORGANIC AND METAL-ORGANIC CHEMISTRY
An International Journal for Rapid Communication in Inorganic and Metal-organic Synthetization

editor: KURT MOEDRITZER, *Monsanto Company, St. Louis, Missouri*

Actively serves the practicing chemist by collecting new methods and techniques for synthetization of compounds, preparation of new and known compounds, and general procedures in organometallic and inorganic chemistry. It makes available for the first time, preparations and procedures that have heretofore been hidden in graduate theses, government reports, and personal communications.

4 issues per volume

SYNTHETIC COMMUNICATIONS

editor: JAMES A. MARSHALL, *Northwestern University, Evanston, Illinois*

In recent years, the use of communications to provide a means for the rapid publication of chemical data has increased markedly. Unfortunately, in those communications pertaining to synthetic organic chemistry, experimental details are usually omitted and generally an extensive wait is needed before they become available. *Synthetic Communications* is a journal in which sufficient experimental detail is presented to permit the reaction or sequence to be repeated by reasonably skilled laboratory workers. The journal assists organic chemists the world over in attaining their goals.

6 issues per volume

C5